Global Oil Finder

Autobiography of a Petroleum Geologist

Fred W. Kelly Jr.

authorHOUSE™

1663 LIBERTY DRIVE, SUITE 200
BLOOMINGTON, INDIANA 47403
(800) 839-8640
WWW.AUTHORHOUSE.COM

First published by AuthorHouse 2/6/2006

ISBN: 1-4259-1901-4 (e)
ISBN: 1-4208-7358-X (sc)
ISBN: 1-4208-7359-8 (dj)

Library of Congress Control Number: 2005906879

Printed in the United States of America
Bloomington, Indiana

This book is printed on acid-free paper.

Table of Contents

Prologue

"Time is like a river made up of the events which happen, and a violent stream; for as soon as a thing has been seen, it is carried away, and another comes in its place, and this will be carried away too."

Marcus Aurelius (121-180 AD)
Philosopher, Roman Emperor

Introduction

Greetings to family, relatives, descendants, friends and anyone else who has picked up this book and will do me the honor of reading it. I am writing this Introduction after passing my 72nd birthday, which I feel is an appropriate time to reflect back on my life. I hope this book will provide an interesting record of the "good old days" and my own early life and adventures. If there are any lessons to be learned here, then I pass them along to you with my good wishes!

It was after my retirement in 1992 that I decided to undertake this book, which has amused me and exercised my mind for the past twelve years. I now deliver this book, however inadequate it may be, to the curiosity and candor of all that read it.

Why have I written this book? Well, in the first place I have always felt sad that my own parents told me almost nothing about their ancestors and their early family life and experiences. I am determined that this will not happen in my case. I am also anxious to give future generations an eyewitness account of what life was like in America during the period 1931 to 1965.

Yet another reason for this autobiography stems from my personal experiences in going through the things of deceased relatives. I have come to realize that boxes of old photos, newspaper clippings, letters, business papers, etc. tend to get thrown out or lost by the wayside over time. I hope that this book will become a permanent, tangible record of my life and times.

For you genealogists out there, I have spent over ten years doing family history research for the first chapter of this book. I belong to the following genealogical societies: Clayton Library Friends, Houston, Texas; Hopkins County Genealogical Society, Madisonville, Kentucky; VA-NC Piedmont Genealogical Society, Danville, Virginia; and the Miramichi Branch of the New Brunswick Genealogical Society, Chatham, New Brunswick, Canada. I hired professionals in Danville, Virginia and Fredericton, New Brunswick, Canada, to help me fill in some of the gaps. All of this has been very rewarding as I have seen my ancestry on both sides of the family unfold in considerable detail back to the 1700s. A historic event like the American Revolutionary

War does not seem all that distant to me anymore, as I now feel that I have an idea about what my great-great-grandparents were doing in those days.

Although it has been a struggle at times to keep going, I have enjoyed writing this book. I have particularly enjoyed re-living past events and re-visiting old friends, relatives and places. I hope my enthusiasm comes through in the pages.

In reflecting back over this book, I am struck by a number of personal revelations, including major social changes, which had not occurred to me before. One revelation is the realization of how several major wars have influenced my life. My maternal grandmother spoke about all the bad things those Yankee soldiers did in the South during the American Civil War. My father fought and was wounded in World War I. I was deeply impressed by World War II as a teenager on the home front and continue to read everything I can find on the subject. And, I have also lived through the Korean War, Vietnam War, and two Persian Gulf Wars. I realize that I have been very fortunate not to have to fight in any wars.

During my lifetime I have been amazed to see numerous major social changes, many as a result of World War II. For example, racial and gender discrimination is now almost a thing of the past and a new multi-cultural society is emerging. Some of these changes have been difficult for me to adjust to because of my "Southern" and conservative upbringing. And, when I was a young geologist I was the lone non-smoker in the smoke-filled conference rooms; I never dreamed that smoking would eventually be banned in all office buildings.

As a scientist and engineer, I have been thrilled by all the new technical advancements that I have seen develop. During my lifetime, old-time vacuum tube radios have evolved into interactive, color, digital televisions with remote controls. The telephone has progressed from asking an operator for a number to rotary dialing to tone dialing to caller-ID and wireless cell-phones with e-mail and color pictures. I have seen the introduction of many other marvelous things now taken for granted such as: helicopters, air-conditioning, video cassette recorders (VCRs), microwave ovens, food freezers, garage door openers, nuclear weapons, nuclear power, space exploration and satellite images. Most people now take television weather satellite images for granted, but I always appreciate that these pictures are coming from weather satellites in geostationary orbit 22,000 miles up in space. All of this was "Buck Rogers" science fiction when I was a boy.

And, then there are today's amazing home computers, which I love to work with. I have been delighted to see them develop from my 1979 Radio Shack TRS-80, Model I (one of the very first home computers with only .004 megabytes of keyboard memory and a tape-recorder for program input and file storage) to today's ultra-fast and powerful desktops, laptops and hand-held wireless devices. I am delighted with today's Internet, e-mail, instant messaging, and other computer activities.

To an old ham radio operator like me, surfing the Internet by computer is the modern equivalent of yesterday's worldwide amateur radio and short wave listening. Radio stations would fade in and out from foreign countries much like today's Internet is made exciting by finding new web sites, flickering videos, and dropped connections.

One of the biggest revelations to me is that my genealogy has revealed that I come from a long line of "free spirits," which left traditions and homes to seek new lands and adventures. My father's ancestors emigrated from Ireland to Canada in 1798, and my father moved from Canada to the U.S. in 1920. Around 1820, my mother's ancestors moved west from Virginia and North Carolina by covered wagon through the Cumberland Gap to Kentucky; then their descendants moved on to Illinois and finally to Missouri, where my mother and I were born. And, I upheld the family tradition by wandering all over the world during my 38-year career with Marathon Oil Company, including living 22 years overseas in Libya, England and Pakistan.

As a boy in our small apartment in Clayton, Missouri, I read the books of Lawrence of Arabia and the adventurer, Richard Halliburton. I dreamed of worldly adventures and was determined to make them come true. I am pleased to report that my resolve resulted in my dreams being realized. You will see in this book that I have: gazed into Mexico's "Well of Death"; lived in tents in the Sahara Desert and drunk tea and spoken Arabic with the local Bedouins and Tuaregs; wandered the bazaars of India and Pakistan; visited the pyramids and royal tombs of Egypt; trekked across East Africa by local bus; and much more. And, although I never went overseas for the money, I also managed to get paid a good salary as a geologist while I was making my dreams come true.

Or, to paraphrase my role model, Richard Halliburton, "writing this autobiography has served to remind me that once upon a time in the far off days of my youth I floated away from the boredom of my school desk by telling myself stories of far horizons, and behold—it all came true!"

I realize now that I never had to worry very much about my personal safety in my vagabond days, especially as Americans were welcomed everywhere. Regrettably, such adventures as mine would appear to be quite dangerous in many parts of the world today due to terrorism and crime.

You will note that during my early adventurous years I often traveled alone and had little interest in social activities. I lived an interesting and entertaining life by myself and with other bachelors until at age 30, during my fourth year in Libya, I got tired of it all and decided it was time to settle down and think about getting married. You will see what a struggle making that change was to be. But, again my determination made it come true.

I was a very lucky man to have fallen in love at first sight with Marcia Grace Mehl, who did me the honor of becoming my wife. Marcia found me a "very rough and rolling stone," but through love, hard work and great patience over almost 40 years she has managed to smooth off some of my roughest edges and make me a more caring person. I am eternally indebted and grateful to her for our wonderful life together, as well as for her patience and help with this autobiography.

I am also greatly indebted to my editor, Margaret A. (Peggy) Stautberg, for her very expert advice and unwavering encouragement in the writing of this book. And, my grateful thanks to my old St. Louis friends, Merle Wolff and Jim Knudstad, for their major contributions to this work. Also, I would like to thank my good friends and fellow geologists, Patrick J. (Pat) Shannon and Harry E. Griffith, for their proficient editing of the technical and geographic portions of this book. And, my cousin, Dana C. Payne, was of immense help in piecing together the genealogy of my mother's side of the family. My grateful thanks to all whom helped and encouraged me.

This volume ends with my marriage to Marcia and the beginning of the second big phase of adventure in my life—namely, the almost 40 years of happy marriage and family life with my loving wife, Marcia, and our three wonderful children, Frederick III (Rick), Heather and Christine. And, now there is our terrific granddaughter, Chelsea; I hope there will be more grandchildren in the future.

I intend to follow up this first autobiography with others covering my life from marriage in 1964 to my retirement in 1992. This work is now in progress but it is an almost overwhelming task to reconstruct the events covering our family activities and my work during nine years in London, six years in Pakistan and 13 years in Houston, Texas.

In order to hedge my bets and whet your appetite for possible additional autobiographies, I have added a final chapter to this book that gives a preview of wondrous and bizarre events in London, England, and Karachi, Pakistan.

God willing and Buffalo Bayou don't rise, I intend to keep writing my story with help from everyone in the family. I only hope it does not take another twelve years!

Thank you for reading this book.

Fred Kelly
Houston, Texas
March 2004

1 ✴

Origins and Ancestry

You are about to travel with me back through time and space to my life experiences during the Great Depression of the 1930s and World War II on the home front, followed by my high school and college dreams of far horizons, and my adventures in the Sahara Desert of Libya searching for oil. You will also get a glimpse of my married life in London, England, and the wilds of Karachi, Pakistan, as I continued my international exploration for oil.

But, first I would like to set the stage for this book by telling you about my family history that undoubtedly had helped to shape my character and world view.

The old, black photostatic copy of my birth certificate records in a bold, Edwardian script that I, Frederick William Kelly, Jr., was born on November 5, 1931, in St. Mary's Hospital in the City of Richmond Heights, St. Louis County, Missouri. My birth certificate also records that my father, Frederick William Kelly, was thirty-seven when I was born while my mother, Dana Ruth (Milleson) Kelly, was only twenty-one. I guess my father followed the old Irish male tradition of marrying late—the sons waited until they inherited the family farm and could support a family—and I was later to carry on the same tradition.

In the summer of 1928 my Mom and Dad met in the romantic setting of Estes Park, Colorado, in the Rocky Mountains northwest of Denver. Dad, a dapper and handsome Canadian bank manager from St. Louis, swept Dana Ruth from Wichita, Kansas, off her feet. They were married in Wichita on June 14, 1930. But, more about that later.

My Mother's Ancestry

My mother's mother, Lela J. (Payne) Milleson, was from a proud southern American heritage. My genealogical research has traced her ancestors all the way back to the 1700s in Virginia and North Carolina. As we shall see, her ancestors include: a Revolutionary War veteran; a veteran of the War of 1812 against the British; a merchant; a Baptist minister; small tobacco farmers; a wealthy tobacco farmer with slaves; a postmaster; and a politician.

Lela's paternal grandmother's family line of Walton goes back to Pittsylvania County, Virginia. Her great-great-grandfather, Lt. Jesse Walton, served in the 1st Georgia Battalion during the American Revolutionary War (so everyone in our family line is eligible to join the Daughters or Sons of the American Revolution); and, he may have been the brother of George Walton, a signer of the Declaration of Independence. Jesse's father, Dr. John Walton, died during the Revolutionary War.

Jesse Walton and his son, William Walton, farmed a plantation in Virginia located about ten miles north of Danville at a place called Pleasant Gap. I know from the old land deeds exactly where the old Walton ancestral home at Pleasant Gap, Virginia was located.

According to the 1820 Federal Census, the Waltons owned seven male and twenty-one female slaves, including fourteen children under fourteen years of age. In those days, an agricultural slave was a very valuable asset, being worth about $30,000 in today's money.

In 1817 William Walton's daughter, Louisa Ann Walton, married Cornelius Payne, who was born in Goochland, Virginia in 1787, and thereby brought the Walton family name into our Payne family tree.

My records show that Cornelius was a veteran of the War of 1812 (1812-15) against British North America (later Canada). Then he became a merchant and ordained minister of the Primitive Baptist Church, while his wife, Louisa, was a governess. Cornelius reportedly built a church in Lynchburg, Virginia.

After the American Revolutionary War ended in 1783, thousands of brave souls traveled west by covered wagon through the Cumberland Gap in the Appalachian Mountains, by way of the Wilderness Road, to territories that later became Kentucky and Tennessee. Cornelius and Louisa decided to join this early migration.

In 1820, Cornelius and Louisa Payne moved west from Virginia to Maury County, Tennessee. The trip would probably have been by covered wagon through the Cumberland Gap and would have taken about six weeks to accomplish. They arrived in Maury County, with its rich farmlands, only 13 years after the county was formed from Indian lands and part of another county. In Maury County, Cornelius served several Baptist churches, taught farming school and farmed.

For some reason, Cornelius and Louisa, and their family of 11 children, left Tennessee in about 1850 and moved north to nearby Logan County, Kentucky. And, then in 1855, they moved to Hopkins County, Kentucky, where they settled in the little town of Nebo. Located about 12 miles west of Madisonville, Nebo had rich farmland and was the center of an important tobacco growing and processing area in western Kentucky.

What was later to become western Kentucky had been real frontier territory in the 1700s. Here the Indian Wars were fought between the British and French over ownership of the land, and the British attacked American settlements during the Revolutionary War. Kentucky finally became a state in 1792, and western Kentucky developed into agricultural lands, mostly tobacco and hemp, using slave labor.

According to a family story, Louisa traveled from Nebo back to Virginia in about 1856 to get money from her father, William Walton, to buy a tobacco farm near Nebo. Her father gave her $600 in gold and she then had a long and rough ride by stagecoach from Virginia back to Kentucky with the bag of gold stuck down her bosom. Louisa, who had a bad disposition (she was known in the family as "that Walton bitch"), complained for the rest of her life that she suffered ill health

CORNELIUS PAYNE (1787-1876)
Primitive Baptist minister
and farmer. My maternal
great-great-grandfather.

LOUISA ANN (WALTON) PAYNE
(1800-1876)
Governess. My maternal great-great-
grandmother.

from the pounding of the gold against her chest during that trip. I have found that the land deeds in Madisonville do indeed show that in 1857 Louisa paid $640 for 128 acres of land on Pond Creek near Nebo. And, she bought it in her own name too!

In 1861 the Civil War began. I have not found any record that any of our southern ancestors served in this war, but some may have. Virginia and North Carolina were, of course, part of the southern Confederacy and fought to preserve state's rights and slavery. Kentucky, however, was tied to the south economically and used slave labor, but was strongly nationalistic, so it tried to remain neutral. But, Kentucky was eventually invaded by northern General Ulysses S. Grant, which forced the State to raise an army to repel him; as a result, Kentuckians served on both sides of the war, which lasted until 1865.

There are two family stories about our Waltons in Virginia during the tragic Civil War. One story is that an ancestor was too ill to fight for the Confederate army and so he sent one of his slaves to fight in his place. Another story asserts that some of our ancestors threw the family silver down their water well to hide it just before Yankee troops occupied their farm during the Civil War.

Cornelius and Louisa Payne both died in Webster County, Kentucky, on the same day in 1876, but I have not been able to find out how they died or where they were buried. Altogether, they had eleven children in Tennessee and Kentucky, including my great-grandfather, John Lee Payne (1839-1899). One of John's sisters had the wonderful southern name of Paralee Tennessee Payne.

John Lee Payne was born in 1839 in Maury County, Tennessee. After his family settled in Nebo, Kentucky in about 1856, he became a farmer, postmaster of Nebo (1895-1899), and for a time served as a member of the Kentucky State Legislature. In 1860, he married Josephine Gooch from nearby Hanson, Kentucky.

Josephine Gooch's father, Willis L. Gooch, was born in Granville County, North Carolina. There are many Goochs in Granville County. When I visited the Court House in Oxford, the county seat, the first thing I saw was a large glass case containing an old funeral coach labeled, "Gooch casket wagon or 'dead wagon' used in the southern part of Granville County by Joseph Henry Gooch (1866-1935)." Joseph was not a relative of ours, as far as I know, but his coach is interesting to see. Gooch is a very ancient and distinguished surname, which originated in the bleak border moors and craggy hills of the Scottish/England border.

I have traced the ancestors of my great-grandmother, Josephine Gooch, back to the mid-1700s in Granville County, North Carolina. The ancestors of Josephine's father, Willis L. Gooch (1807-1842), were his father, Thomas Gooch (1784-1859), his father, Rowland Gooch (1762-1822), and his father, John Gooch (about 1740-1793). One census document indicates they were from the Knap of Reeds area near Butner in the extreme southwestern corner of Granville County.

In the 1820s, Willis L. Gooch and his wife, Rachel (Cozart) Gooch, moved west from North Carolina to Hanson, Kentucky, located just north of Nebo. They probably traveled by covered wagon through the Cumberland Gap to Hanson.

According to my Aunt Lucille (Milleson) Perrings, our Gooch family goes back to colonial Williamsburg and to Sir William Gooch, the British colonial governor of Virginia from 1727 to 1749. I have not yet established a direct link between our Gooch family and Governor Gooch, but the gap is only a couple of generations wide. Perhaps we are what some call "collateral descendants" of Governor Gooch, which is, I guess, a nice way of saying that we may be somehow related to the old governor.

John Lee Payne and his wife, Josephine, had ten children in Nebo, Kentucky including my grandmother, Lela J. Payne, who was born in 1869. Lela was born just four years after the Civil War had ended, and I remember her telling the stories she had heard as a little girl about all the bad things those "Yankee" soldiers had done to the people in the South.

Lela Payne married John Franklin Milleson in Nebo, Kentucky, in 1893. Perhaps they met in Nebo when John, a Yankee from Iowa, passed through town plying his hardware trade. The couple established their home in Nebo and daughters Lucille and Kathryn (Peg) Milleson were born there. At this time, Nebo was flourishing as it was the third largest tobacco stripping market in the world and had up to seven tobacco factories operating. In 1875, a railroad had reached the town to haul away tobacco and coal from the nearby mines. There were good schools and churches, and an active social life.

My mother's father, John Franklin Milleson, was born in Russell, Iowa, in 1874 and eventually got into the hardware business. I do not remember anything about him because I was only three years old when he died in 1934. John played the coronet for a hobby and lost the vision in one eye to glaucoma (which runs in our family).

According to my Aunt Lucille, John's family was of French and Welch origins. However, recent information that I obtained from the Internet indicates that the first Millesons arrived in America from Scotland in the early 1700s.

Marriage records show that our John Milleson's father was born in Illinois and his mother was born in Indiana, but that is all I have been able to find out about his family origins. According to my cousin, Bob Johnson, my grandfather was a very kindly man. When Bob was a boy, John taught him to play the coronet and showed him things like how wheat grows and is harvested.

In about 1899, John and Lela moved with their two daughters to Payna, Illinois, where their third daughter, Sybil Milleson, was born. Then they moved to St. Louis, Missouri as John pursued his hardware trade. My mother, Dana Ruth Milleson, was born in 1910 while they lived in St. Louis.

Grandfather John owned a hardware business in St. Louis and used to make frequent trips by train to Mexico to trade hardware goods.

JOHN FRANKLIN MILLESON (1864-1934)
My maternal grandfather, a hardware manufacturer, at
age 33 years in about 1897 in Ottumwa, Iowa.

JOHN AND LELA MILLESON IN 1909.
My maternal grandfather and grandmother
vacationing in Eureka Springs, Arkansas.

On one long trip to Mexico he left his brother in charge of the business and when he returned the business had gone bankrupt. Grandmother Lela always said John was not any good with money, and I guess the same can be said about his brother!

After the hardware business failed in the 1920s, John and Lela had no money and their marriage fell apart. There was no Social Security "safety net" in those days, so the penniless Millesons turned to relatives for support. They left St. Louis and moved to Wichita, Kansas, to be near my mother's two older married sisters, Sybil and Kathryn ("Peg"). Sybil, a professional singer, and her family lived in Wichita, while Peg and her husband lived in Independence, Kansas. Later, John moved to St. Louis to live with their daughter, Lucille Perrings, and her family.

Grandfather John died suddenly in 1934 at age sixty while living in St. Louis with my Aunt Lucille Perrings and her family. My cousin, Pat Perrings, clearly remembers that John collapsed and died while they were playing in the street near her house. He was buried at Sunset Memorial Park in St. Louis.

Grandmother Lela moved around staying with all her daughters. She briefly lived with Mother and our family in our small apartment in Clayton, Missouri; so, I remember her quite well. She died in 1943 at age seventy-three in a nursing home outside St. Louis. I vividly recall visiting her in that very depressing and disturbing home although I was only eleven years old at the time. Lela is buried at Sunset Memorial Park in St. Louis next to her husband, John.

Great Aunt Goochye

When I think of my grandmother, Lela, I often recall all the fascinating stories I heard in our family about her younger sister, Ethel Goochye Payne, known as "Goochye" or "Goochie." I guess her unusual middle name was a diminutive form of her mother's family name of Gooch. Anyway, my Great Aunt Goochye married Dana Charles King (a distant cousin of Admiral Ernest King of World War II fame, and his brother, Dr. Glen King, who was the first to isolate Vitamin C) in 1904. They moved to Los Angeles, California, where Dana became one of the founders of the giant Sunkist orange co-operative. For many years he was president of Sunkist's shipping division, and he and Aunt

LELA J. (PAYNE) MILLESON
(1869-1950)
My maternal grandmother in a photograph taken about 1930.

Goochye were our only really wealthy relatives. They lived in a very large house in the prestigious Bel Air neighborhood of Los Angeles, located at 227 Copa De Oro Road. Their neighbors at one time were the actor Dick Powell and his wife, June Allison.

In addition to their Bel Air mansion, Uncle Dana had a 200-acre ranch of navel oranges, known as the King Ranch. Meanwhile, Aunt Goochye had her own ranch, which consisted of 10 acres of lemon trees. Uncle Dana and Goochye competed every year to see who would have the best crop, and frequently the lemons won.

Just before the Great Depression started with the Wall Street Crash of October 29, 1929, Sunkist was shipping its first oranges to England in refrigerated ships. Uncle Dana and Goochye planned to make a trip on one of the refrigerator ships to England. But, Uncle Dana was concerned that he would be out of touch with the stock market while he was at sea and so he sold all his stocks just before they sailed. While they were on their trip, the stock market crashed but all of Uncle Dana's money was in banks, and none of his banks failed during the Depression.

GOOCHYE AND DANA KING
On their ranch in California (1930s?).

Also, before making their trip to England, Aunt Goochye bought an expensive fur coat for the trip. While they were on their trip to England, the stock market crash and start of the Depression caused the value of her fur coat to fall drastically. When Aunt Goochye returned to Los Angeles, she took the fur coat back to the store, where she was a very good customer, and demanded the original cost of the coat back; and, she got it!

Aunt Goochye died a widow in 1961. Having no children, she left practically all of her estate to take care of her invalid sister, Betty, and to charities. In addition, Aunt Goochye felt that every woman should have some money of her own. So, she specified in her will that an equal—but relatively small—sum of money should go to each of the woman in the family. This included my mother and her sisters, Peg, Sybil, and Lucille, and Lucille's daughters, Ginny and Pat. This action caused much consternation to my mother and her two sisters that did not have daughters, because my brother, Jack, and I, and our cousin, Bob Johnson, did not receive any of Aunt Goochye's money. As a result, the three sisters that did not have daughters got angry with the one that had two. I guess it all proved the old saying that, "If you want to really get to know people, share an inheritance with them!"

My Mother

My mother, Dana Ruth (Milleson) Kelly, was born on May 17, 1910, in St. Louis, Missouri, while her father's hardware business was doing well. However, Mother was forced to move to Wichita, Kansas, with her mother and father sometime in the 1920s after her father's hardware business failed. They then had to look to mother's sister, Sybil, for a place to live, with financial help from her other sisters.

Mother graduated from Wichita High School in 1927 and then attended Fairmount College in Wichita (now Wichita University) for a year or two. She belonged to the Alpha Tau Sigma sorority and became a very popular member of Wichita society. After being with Sybil and her family for several years, Mother left college and moved to Independence, Kansas, to live with her sister, Kathryn (Peg), and her oilman husband, Marian A. Halsey. Marian's family owned the local dry goods store and was socially prominent in the small town of Independence.

Peg and Marian Halsey had no children and cared for my mother almost like a daughter when she was a young lady. And, ironically, twenty years later the Halseys were to put me through college almost like a son.

Mother became a kindergarten teacher in Independence and a very active member of the social scene in both Independence and Wichita. She attended bridge parties and country club affairs and also played the piano. In the summer of 1928 the Halseys took Mother to Estes Park in the Rocky Mountains of Colorado, their favorite place to escape the hot summers of Independence, Kansas. At Elkhorn Lodge in Estes Park, Mother met the man who would become her husband and my father, Frederick William Kelly, a bank manager vacationing from St. Louis, Missouri.

My mother was a very pretty lady; to get a good idea of what she looked like just watch any old movie from the 1930s or 1940s starring Barbara Stanwyck as I have always thought they were look-alikes. Mother was also very friendly, outgoing, and well liked. When I was in grade school, she took part in community affairs in St. Louis, such as counting votes in elections, but later had to give this up when she had to go to work. Every Christmas she made gifts for her friends such as white "divinity" fudge, dates stuffed with pecans and rolled in powdered sugar, and Spirea branches dipped in glue and imitation snow flakes which were used for decorations. I well remember as a small boy stuffing and powdering dates and driving out to the country with her to gather branches of Spirea.

I feel that I inherited my active and imaginative mind from my mother. I also inherited pride in her southern family roots and in the Confederate South. I have visited the Gettysburg Civil War Battlefield in Pennsylvania several times and always feel the sorrow of the bloody Confederate defeat there.

As will be seen in later chapters, Mom started married life as a society lady but following The Great Depression became a working and worrying mother. A heavy smoker, she died on April 5, 1982 from emphysema and was buried next to Dad in Sunset Memorial Park in St. Louis.

My Father's Ancestry

My father, Frederick William Kelly, was born in Montreal, Quebec, Canada, on October 31, 1891, and he was a fourth generation Canadian.

My father's great-grandfather, James M. Kelly, was probably born in the Belfast area of northern Ireland sometime in the late 1700s; but we do not know anything about his family in Ireland. We do know that he emigrated from Belfast to New Brunswick, Canada (then known as British North America) in 1798, according to several published biographies. We do not know why James left Ireland in 1798, but it might have had something to do with the violence associated with the Irish Rebellion of 1798 against the British (which failed). However, we do know something about the emigration patterns going on at that time thanks to an excellent book by Houston and Smyth entitled, *Irish Emigration and Canadian Settlement* (University of Toronto Press, 1990).

Following the defeat of the French in North America by the British in 1759-60, the British authorities needed loyal subjects to secure their new territories, including New Brunswick. Many new emigrants came from northern Ireland because it was the closest part of Britain to Canada and therefore provided the lowest fares across the Atlantic. Migrations in the late 1700s proceeded the time of regular steamship travel. Our James probably would have arrived in Canada on a sailing ship that had carried lumber, coal, slate or other bulk goods to Europe. Emigrants, such as James, were in effect moneymaking ballast when these ships returned to Canada.

Most of the earlier Irish emigrants to Canada were Protestants, as was James. The high cost of passage singled out relatively higher class emigrants, mostly Protestants of northern Ireland. These emigrants had enough money to see them through a journey that few Irish had yet attempted. In other words, Irish emigrants of the late 1700's and early 1800's were from the better off, and only later would poorer people join the emigrant flow.

James's voyage to Canada could have lasted anywhere from six to ten weeks, during which time he would have faced the dangers of fire, disease, storms and icebergs, to say nothing of chronic seasickness. The sight of land must have been very welcome. He probably disembarked on the docks of Saint John, New Brunswick, in the summer or early fall of 1798 and would have found his new surrounding very different from Belfast. We know that he then settled in the nearby town of Bend of the Petitcodiac, now known as Moncton, New Brunswick.

Records show that James owned a public house, or pub, on King Street in Moncton in 1827. And, so he was not a farmer, which might say something about his family situation back in Ireland. James and his family lived on their own land near Moncton, but he apparently speculated in land sales and had financial problems that led to all his assets being sold at auction by the courts in 1838. A very public notice of the auction appeared in the newspapers!

James's first wife died in 1836, but he re-married in late 1838 to Mary E. Chapman, the daughter of the Sub-Collector of H.M. Customs in nearby Dorchester, New Brunswick. His prosperity appears to have improved considerably following his re-marriage.

We are fortunate that in 1827 my great-great-grandfather, James M. Kelly, had a son, William Moore Kelly, who became a prominent and well-liked politician and businessman in New Brunswick. As a result, there are two published biographies about William that also give some information about his father, James. These two excellent biographies can be found in the *Dictionary of Canadian Biography* (University of Toronto Press, 1998) and the *Dictionary of Miramichi Biography* (Willis D. Hamilton, 1997).

According to biographies, William and his father, James M. Kelly, moved from Moncton north to the Miramichi (pronounced "meer-ma-SHEE") River Region of New Brunswick in about 1839. In that year, James obtained a contract from the New Brunswick provincial government to operate a mail stagecoach service once a week between the Miramichi and Fredericton, the provincial capital. The Miramachi Region of New Brunswick centers around the Miramichi River, which is world-famous for its salmon fishing. The Miramichi River cuts through the heart of New Brunswick and empties on the East Coast into the Gulf of St. Lawrence, which connects to the North Atlantic Ocean. Straddling the Miramichi River, near its mouth, are the twin cities of Chatham and Newcastle. These cities had been world centers of lumber and shipbuilding since the 1700s. And, more importantly to us, it was in these cities that our Kelly and Fraser ancestors were living and working in the mid-1800s.

Sometime after 1839, James added the Miramichi to Moncton route to his growing stagecoach business, and he carried both mail and passengers. James is considered to be real pioneer of New Brunswick, as he arrived less than 40 years after the "British Conquest" of French Canada in 1759-60. He preceded even the first major Irish influx during 1820-30, and arrived almost 50 years before the rush of Irish emigrants during the Great Irish Famine of 1845-50.

James died in 1844 in Newcastle, Miramichi, New Brunswick. His eldest son, William Moore Kelly, succeeded him in the stagecoach business, although he was still less than eighteen years of age at the time. William (my great-grandfather) was born in 1827 while his father was running a pub in Moncton, New Brunswick. In 1847, he married Eliza Ann Long, daughter of James Long of Cocagne, New Brunswick.

William inherited his father's stagecoach business, but he was not a stagecoach driver. His younger brothers Joseph and Robert Kelly were among four stage drivers being employed by him in New Brunswick in 1851. Later in the 1850's he was a partner in the Fredericton to Miramichi stage route. In 1860 William started a daily mail and passenger service between Chatham and Shediac, which had become the terminus of the Eastern and North American Railway line from Saint John, New Brunswick. He was still a carrying the mail in 1871.

One biography amusingly reports that on one occasion a Kelly stage also pulled a load of lumber and it struggled up the hills and then raced down the other side at breakneck speeds. The driver was said to have been fast asleep and to have only awakened when the coach was heading for the cliffs on the side of the road!

William Moore Kelly took over the Royal Hotel in Chatham in 1849 to service his stagecoaches and he renamed it Miramichi House. He soon got out of the hotel business, however, but continued to make Chatham his hometown.

In 1856, William's first wife, Elizabeth (Long) Kelly, died, and in 1857 he married Margaret Fraser. Margaret was the second daughter of Alexander Fraser, Jr., and his wife, Catherine (Fraser) Fraser, both of whom had emigrated to Canada in about 1804 from Inverness-shire, the home of the Fraser Clan in Scotland. Alexander was from Stratherrick and Catherine was from Gorthlick in Inverness-shire.

Alexander Fraser, Jr., known as "Long Fraser" to distinguish him from others of the same name, was a prominent businessman in Chatham. He owned a general store and a giant steam-powered sawmill, and was also a Justice of the Peace and a commanding officer of the Chatham Rifles militia brigade. Margaret Fraser's marriage to William Moore brought the Fraser Clan into our Kelly family line.

William Moore Kelly entered provincial politics in 1867 when he was nominated for a seat in the New Brunswick House of Assembly. He took the seat by acclamation and was also successful in elections held in 1869, 1870, and 1874. In 1869 he was appointed chief commissioner of public works and a number of the first "great bridges" of New Brunswick dated from his years in office.

William was a good and popular minister, but he became involved in contentious political debates because he was a Conservative. He supported confederation of the British North American colonies into a united Canada, which at the time was widely opposed in New Brunswick because of fears that this province would be overshadowed by the larger provinces and not be given effective voting power in a union parliament (an agreement for Confederation was finally negotiated and passed in 1866). William also supported free education, direct taxation for the maintenance of schools, and the construction of an interprovincial railroad. Politics got pretty rough in New Brunswick during this period, including fist fights in the Parliament and much "mud-slinging" (one local Chatham politician who verbally attacked William had the grand name of Jabez Bunting Snowball!).

William became very involved in a railroad being constructed in New Brunswick as both a government official, who dispensed subsidies, and a private investor. In 1871 he was criticized for accepting railroad mail contracts while a member of the government, and he was then censured for granting rail subsidies in areas outside his own constituency. Finally, although he had a good reputation, he was accused by rivals of granting government favors to railroads in which he had a personal interest. Having had enough of all these altercations, he retired to the less controversial New Brunswick Legislative Council in 1878.

Suffering from ill health, William Moore Kelly and his family moved to Toronto, Ontario, Canada, in 1882, without resigning his seat on the Legislative Council. According to one biography, "business reverses came upon him" in his last years, and in the end "his mental powers gave way." William's death occurred on December 12, 1888 at the Montreal home of one of his sons, Frederick Fraser Kelly. His remains were returned to Chatham for burial in the cemetery of St. Paul's Anglican Church at Chatham Head, Miramichi. I visited this cemetery in August 2001, and was very impressed and moved by the six-foot high, white, marble obelisk that marks his grave.

William Moore Kelly had at least seven children who survived infancy. These were a daughter of his marriage to Elizabeth Long, along with four sons and two daughters from his marriage to Margaret Fraser. One of his sons with Margaret Fraser was Frederick Fraser Kelly, my father's father and my grandfather.

Frederick Fraser Kelly was born in Chatham, New Brunswick in about 1858. In about 1890 he moved to Montreal, Quebec, Canada, and married Lucy Jellett, whose father was Judge Robert Patterson Jellett and mother was Charlotte MacNider Jellett. Frederick took his bride Lucy on a sailing ship cruise down the East Coast of America on their honeymoon.

LUCY JELLETT AND HER MOTHER IN 1865.
My great-great-grandmother, Lucy MacNider
Jellett, and her mother, Catherine MacNider, in
a photograph taken about 1865 in Canada.

Lucy's father, Judge Jellett, was born in Belfast, Ireland, and graduated from the University of Dublin in Ireland. He emigrated to Canada and became a very prominent judge in the Belleville-Picton area of Ontario, Canada. Jellett is a very distinguished old Irish family. According to a family story, one of our Jellett ancestors was a British general in the Boer War (1899-1902) in South Africa and was known as the "Savior of Ladysmith" after he marched his army into the town of Ladysmith and saved the British Garrison from a Boer attack.

Frederick Fraser and Lucy lived at 46 Tupper Street in Montreal when my father, Frederick William Kelly, was born in 1891. I visited that street in 1998 and found it to be lined with well-built, stone, Victorian-style townhouses looking just as they probably looked in 1891. The houses have been renumbered but based on a few old gold

LUCY (JELLETT) KELLY
(1861-1931)
My Canadian paternal grandmother.

FREDERICK FRASER KELLY
(1858?-1938?)
My Canadian paternal grandfather.

numbers I found, and information from a cousin, the new number on the old Kelly home is believed to be 1922 Tupper Street.

Frederick Fraser Kelly was in partnership with C.B. Kelly (his brother Carling?) in the import and wholesale business in Montreal. He and his family moved to Toronto and in 1909 he was a "traveler" for a large company and in 1912-13 was listed in the Toronto phone books as a "manufacturer's agent." According to my Aunt Gretchen (Kelly) Ballantyne he was a salesman for the Gordon MacKay Company at some time around 1900. He died in Toronto in about 1936 and I do not recall ever meeting him. His wife, Lucy, died before him in 1931, the year I was born. And so, unfortunately, I do not recall anything about my grandparents on my father's side.

My Father

My father, Frederick William Kelly, was born in Montreal, Quebec, Canada, on October 31, 1891, and he was known in the family as "Fritz." (For some reason, the family had several German first names and nicknames about this time, such as Fritz, Gretchen, and Greta.)

The blackened molds were clamped together, and we waited impatiently for a bluish scum to form on the molten lead, which meant it was hot enough to pour. Then, I carefully poured the lead into the mold in a steady stream. After a brief wait, Jack pulled the mold out of the clamp and separated the two halves.

Opening the mold was the moment of truth. We were usually thrilled to find three complete, shiny, lead soldiers lying in the black smoky mold. Sometimes, however, an arm or leg was missing, so the pieces went back into the pot for another try.

We arranged our new soldiers into a battle scene with toy tanks and planes and then got out our ElecToy electric cannon. Jack dropped a wooden bullet down the black, Bakelite barrel of the cannon. I aimed and set the angle of the barrel, and then pressed the electric button to fire the gun. The electromagnet in the black box at the base of the barrel made a loud bang and shot the wooden projectile across the room into the ranks of shiny lead soldiers. After a long bombardment, often finished off with a BB gun barrage, we again had a pile of broken lead soldiers ready to go into the melting pot on another rainy day.

[Fifty years later, I still get my home casting set out occasionally just to show the neighborhood kids how we had fun in the "good old days." Can anyone imagine a manufacturer selling a toy to kids these days that involves pouring molten lead into aluminum molds that have been smoked over an open flame? When I was a boy, there was no thought about lawsuits; if you got hurt it was your own tough luck!]

Every now and then our Aunt Cile (Lucille) Perrings would come by in her big Buick and take Mother, Jack and I out into the country to the farm of Uncle Bob and Aunt Hattie Payne. Bob was actually mother's uncle (her mother's brother), which would have made him my great-uncle. Uncle Bob's farm did not have electricity, running water or indoor plumbing! Kerosene lamps provided light, wood provided heat for cooking and warmth, batteries ran the radio, and a hand pump in the yard provided water. Some distance away from the house there was an outdoor toilet, known as an "outhouse" or "privy." The privy was a small wooden hut with two holes cut in a board as toilet seats; when the smell got too bad, powdered lime was shoveled through the

Dad told me once that he had been named after the Prussian (German) King Frederick William II, who ruled from 1740 to 1786 and was known as "Frederick The Great," and later in his life as "Old Fritz."

I do not know much about my father's younger years. I have no record of his life in Montreal except that he and his family lived at 46 Tupper Street. Then, Dad and his family moved to Toronto, Ontario in about 1909 when he was about 18 years old. The 1912-13 Toronto telephone book shows Dad as working as a clerk for the Bank of Toronto. He enjoyed sailing, skiing, ice hockey and golf. A family photo album shows him in his twenties sailing on Ontario lakes with his sisters and friends while wearing a white shirt and tie with sleeves rolled up.

Candid photos of Dad at about age 23 show him with seven other men at a tent camp-out at Pushlinch Lake in Ontario with everyone drinking beer and playing cards. One of the men wrote on a photo of Dad that he was, "Cook Kelly on whom devolved the work of getting dinner. The cook is most genial although not interested in pitching the tents."

World War I started in Europe in 1914, with Great Britain and other Western European powers (and later the United States) fighting a war started by Germany, Austria-Hungary and Turkey. When Canada became involved, my father trained at the Royal School of Artillery in Kingston, Ontario, and was commissioned as a lieutenant on March 10, 1917.

Dad sailed to France and commanded a trench mortar battery in the Canadian Expeditionary Force attached to the British Army. He was wounded by German shrapnel on September 3, 1918, probably in the Somme River area of Picardy, France. He spent some time at a hospital in London, England, where they removed some but not all of the shrapnel in his left hand, right shoulder, and right back and thigh. He was then demobilized on February 26, 1918, some months before the end of the war, but carried shrapnel in his body for the rest of his life. He was lucky to have even survived the bloody war at all as almost one-fourth of all the Canadian troops in the war zone were killed (56,000 killed).

Dad's .45 caliber, Colt Automatic service pistol and war medals (probably the silver British War Medal and the bronze Canadian

FREDERICK WILLIAM KELLY (SR.)
My father as a lieutenant in the Canadian
Army in 1917 preparing to go to France
to fight in World War I.

Victory Medal 1914-1918) were in the family when I was a boy but have since been lost. However, I have managed to get all his military records from the Government of Canada.

Dad immigrated to the United States from Canada in 1920 but retained his Canadian citizenship. He eventually became manager of the First National Old Colony Bank in St. Louis, Missouri. He told me once that before he married my mother, he had lived at the posh University Club in St. Louis, where he told them he had "graduated from the Royal University in Kingston, Ontario." In fact, the "Royal University" was his old army artillery school and he never went to college.

My Dad was a very dapper and nice-looking man. He wore a small, closely trimmed mustache and dressed very neatly. I have always thought that Dad was very "British" in his manner and dress, as might be expected of a Canadian of his era. He was a very quiet man who very much kept his thoughts to himself. In fact, I find it hard to remember his ever sharing many of his thoughts with me.

Dad was a very religious man, although he rarely attended his Episcopalian Church because Mother was a Presbyterian; apparently Dad agreed to let Mom raise the family as Presbyterians but he did not participate. But, I can still remember Dad and Mom praying for a job when he was out of work while I was in high school. Dad was also very prudish, and no nudity or discussion about sex was allowed in our household, as was common during this rather puritanical period of family life. He also censored the radio adventure stories we were allowed to listen to if he thought they were too violent. But Dad did have a dry, British sense of humor and loved to tell a few jokes and play practical jokes.

As will be seen in subsequent chapters, Dad lost his job as a bank manager and everything else during the Great Depression of the 1930's, and in many ways he never really recovered from this during the rest of his life. He suffered a cerebral stroke in 1960, which severely incapacitated him and resulted in my brother, Jack, staying home to take care of him full-time instead of working (and Jack did a wonderful job of caring for Dad). Dad suffered another stroke and then died at age seventy-one in St. Mary's Hospital in St. Louis—the same hospital in which I was born—on December 9, 1962, while I was working in Libya. Dad was buried in Sunset Memorial Park in St. Louis.

I feel that I have inherited my appearance, rather introverted outlook, and British sense of humor from my father—also his full head of hair for a lifetime, as my brother, Jack, who has the same head of hair, proudly reminds me.

My Parents' Wedding and Honeymoon

In the summer of 1928, Dad (Frederick William Kelly) went on vacation to Estes Park, Colorado, where he met Dana Ruth Milleson, who was to become his wife and my mother. After meeting in Estes Park, Frederick went back to his bank in St. Louis and Dana Ruth returned to teaching in Independence, Kansas. Then, in the fall of 1929, Dana Ruth went to St. Louis to become a junior at Washington University. While there, she got back together with Frederick. On New Year's Eve, they drove to Jefferson City, Missouri, to celebrate with friends. On the way back to St. Louis on New Year's Day 1930, they became engaged to be married.

Mother and Dad were married on June 14, 1930, in Wichita, Kansas, where Mother had many relatives and friends. The wedding was held at the St. James Episcopal Church with an Episcopal priest officiating on behalf of Dad's church and a Presbyterian minister representing Mother's church. The wedding was the social event of the season with many large articles and photographs in the Wichita newspapers. The wedding was described as "a nuptial ceremony of charm," and Mother was described as "a bride who possesses enviable charm and beauty." A local newspaper article some weeks later discussed whether or not it was appropriate for the groom to kiss the bride at weddings; it noted that at the Frederick Kelly-Dana Ruth Milleson wedding "the altar kiss was used to complete the double ring ceremony."

Mother and Dad, as bride and groom, left the church for the wedding reception on the back of an open truck loaded with refrigerators and trailing tin cans. This was all a gag set up by Mother's brother-in-law, Carl Johnson, who was in the electrical appliance business and quite a joker. My Uncle Carl also took very early 16-mm. home movies of their departure from the church by truck. For many years when we were boys, my brother, Jack, and I watched that wedding movie on our toy 16 mm. projector when it was raining and we could not play outside; sadly, that piece of film has been lost.

After the reception at Sybil and Carl Johnson's house in Wichita, the bride and groom, according to a local newspaper account, "departed by train for St. Louis, from where they will motor in his open car to Toronto, Canada, to be guests of the bridegroom's parents, Mr. and Mrs. F.F. Kelly. From there they will go to Montreal to visit the bridegroom's brothers, continue across Quebec, and from there travel down the Atlantic Coast. They will be at home in St. Louis after July 14." They stopped in New York City near the end of their honeymoon and then headed for St. Louis to set up their new home together.

FREDERICK WILLIAM KELLY (SR.) (1891-1962)
My father in a photograph taken in 1930.

DANA RUTH (MILLESON) KELLY (1910-1982)
My mother in a photograph taken in 1930.

Dad was manager of a bank, so the newly married couple was financially well off and enjoyed St. Louis society. They moved into a furnished apartment in Clayton (15 South Lyle Avenue) while they decided what furniture to buy and where to live. In August 1930 they bought their furniture [I have inherited several pieces of this furniture] and moved into a large, unfurnished apartment a block away at 7533 Buckingham Drive. Those apartment buildings are still there today and largely unchanged. It is amazing to see the same ornate stone work on the buildings today that is identical to that seen in photos of me when I was a baby!

According to Mother's bridal book, she and Dad celebrated their first wedding anniversary in St. Louis with dinner at the Westborough Country Club followed by a musical show at the open-air Municipal Opera in Forest Park. They celebrated their second anniversary with a group of friends; they had "high balls" (cocktails) at the home of some friends, dinner at a prominent restaurant, and then had a "grand time" at the Meadowbrook Country Club.

My Arrival

I was born on November 5, 1931. I guess I am a so-called "Depression baby" as the Great Depression had been going on since the October 29, 1929, stock market crash. As a result, I have been very conservative and security conscious all my life as is said to be typical of Depression babies. My brother, John Fraser Kelly, or "Jack," was born two years later on September 30, 1933.

In spite of the Depression, our family was financially comfortable when Jack and I were little boys. Dad was manager of the First National Old Colony Bank in St. Louis and had also received a small inheritance some years before. In 1933 we moved to another nearby apartment at 7515 Parkdale in Clayton and then to a nice, two-story brick house at 515 North Central in adjoining University City, an affluent subdivision that includes Washington University.

My earliest recollections are of myself as a small boy about three or four years old playing in the backyard of our house on North Central. This house still looks today almost exactly as it did when I was a very little boy. According to family photographs, we had a big Cadillac car. And, Mother wore furs and belonged to the socially prominent Junior League. So, we were doing quite well while most of the country was deep in the poverty of the Great Depression.

2 ✳

A Boy's Life

In the summer of 1936, when I was almost five years old, our family traveled by car to Ontario, Canada, to visit my father's Canadian relatives and I remember the trip very clearly. We drove from St. Louis to the spectacular Niagara Falls—which I remembered the rest of my life and prompted me to revisit in 1998—located on the border between New York State and Ontario, Canada. After a brief stay at Niagara Falls, we drove on to Toronto, Ontario, where most of my Canadian relatives were living. After a nice visit in Toronto, my family and I, together with our Canadian relatives, drove about 100 miles northwest to Georgian Bay, an arm of Lake Huron. We then took a ferryboat out to a tiny island owned by Dad's sister, Gretchen Ballantyne, and her family.

The rocky little island had a nice, comfortable lake house with a big front porch looking out on the water. There was also a boathouse with walls hung with life preservers and a boat jetty where my brother, Jack, and I tried to catch minnows with little nets. Jack and I spent most of our time on a little, muddy beach, where we swam and made toys and dishes out of mud. The water was so clean it was pumped up into the house and we drank it right out of the lake! I also recall that Mother got her feet terribly sunburned while sitting out on the porch. It was a fun vacation that I will never forget, even though I was not quite five years old at the time.

MOTHER, JACK AND I IN 1934.
That's me on the left at about three years old in St. Louis, Missouri.

In September 1936, I started elementary school at the Flynn Park School in the affluent community of University City, Missouri, near Washington University. I walked to school with Mother in all kinds of weather with my peanut butter-and-jelly sandwiches. One Christmas morning at about this time Jack and I were both wrapped in blankets and carried down-stairs to the Christmas tree, as we were both sick with some childhood malady.

There was no air-conditioning in those days, so, when winter passed, one thing we did to cool off on hot, humid evenings was to drive to the Pevely Dairy in nearby Clayton. Pevely had a big fountain with sprays that changed height and color. Cars parked in concentric circles around the fountain and young girls served ice cream to the customers. I can still visualize the cool spray from the fountain drifting over Jack and me as we sat on the curb eating ice cream!

As previously mentioned, the Great Depression had been going on since the Stock Market Crash of October 29, 1929, known as "Black Friday." The Crash marked the end of the "Roaring Twenties," which had been a post-World War I period of prosperity and over-indulgence. The economy collapsed, farmers lost their land, many lost their life savings, and millions of men were out of work.

But, in the early years of the Depression my family was doing well with Dad managing a bank. However, the "good life" ended for our family sometime in the middle of the 1930's. Dad lost his job when his bank (and thousands of others) finally failed due to the Depression. By 1938 there were over eight million men out of work, or about 20 percent of the work force. Things were tough!

In 1940, we had to leave our nice house at 515 North Central in University City and move to a small rented apartment in the adjacent suburb of Clayton, Missouri. Dad got a job with the Federal Reserve Bank in downtown St. Louis. After we moved to Clayton, I changed from Flynn Park School to Meramec Elementary School, very close to our apartment at 216 South Meramec Avenue. Meramec School was built for affluent neighborhoods on the other side of the railroad tracks from our apartments (parents from our side had to fight to get us into the school) and was very modern and well equipped. It had wondrous new things like white boards with black chalk and room lights that automatically came on when it got dark; it was also one of the first American schools to play soccer for physical education. Meramec School was one of the first so-called "progressive" schools in the country. Educators came from all over the United States to watch us as "each student worked at his or her own pace," and with no homework. Mother blamed this radical new system for many of my later school problems in reading, English and mathematics.

Although I was not aware of it, the Chancellor of Germany, Adolf Hitler, known as *Der Furhrer*," and his Nazi Party had formed the fascist *Third Reich* and then started World War II by invading Poland in 1939. And Germany's ally, Japan, invaded China. But, the United States remained neutral and stayed out of the war, at least for the time being. However, on December 7, 1941, Japanese airplanes from aircraft carriers launched a surprise attack on our naval forces at Pearl Harbor in Hawaii without even declaring war first. Japan then declared war

JACK AND I FISHING IN 1941.
That's me on the left at about 10 years old fishing in northeastern Oklahoma
on vacation with our Aunt Peg and Uncle Marian Halsey of Tulsa, Oklahoma.

on us and Japan's allies, Germany and Italy, followed by declaring war on us, too. World War II had started for the United States. The country now went on a major wartime effort that was to last for five years and cost over 400,000 American lives. The only positive note about going to war was that it ended the Depression and put everyone back to work.

I was 10 years old when Pearl Harbor was attacked, but cannot remember what I was doing when the attack was announced. What I do remember from this time was that Dad soon lost his job with the Federal Reserve Bank because of a new wartime law that aliens could not hold government jobs. Although Dad had been in the United States since he left Canada in 1920, he had refused to apply for American citizenship. Dad hunted for another job for some months and finally got one with a small stock brokerage firm in downtown St. Louis. In the evening, he sat on the sofa in our little living room and went over the day's stock market results. But, for some reason this job did not last very long either. I still vividly recall going downtown alone to his office to collect his personal effects. It was very disturbing for me to have some man I didn't know point over to a desk and say, "your Dad's things are over there."

Dad then worked for the St. Louis electric power company, Union Electric, for a brief time, and following that he tried to sell Martha Washington brand candy door-to-door (the samples were a treat for us). Meanwhile, our car was sold, and so we did not have a car all through my high school years. Mother and Dad would kneel down and pray that Dad would find a permanent job soon, which made Jack and I worry all the more about the situation. Hoping it would help get him a job, my father finally decided to give up his Canadian citizenship, as my mother had wanted him to do for years. He became a naturalized American citizen on November 6, 1942, at 51 years of age.

Finally Dad got a permanent job as an accountant at a fuel yard some miles away from our apartment. He left home six days a week before sun-up in the morning to catch a bus to work and then got home dead-tired after dark. (This job was to go on throughout my high school years, too, and as a result I saw very little of Dad; I guess he was what is now called an "absent father.")

World War II lasted from 1941 to 1945 and made a big and lasting impression on me. I was 10 to 14 years old during the war and so was too young to serve in it, but I well remember "life on the home front." These were very exciting days for me! We made wooden model warplanes in Boy Scouts and painted them black for our soldiers to use to identify enemy aircraft. We lived on ration stamps and ration tokens for scarce items, such as: sugar, coffee, meat, shoes, alcoholic drinks, and gasoline (but we didn't have a car). Ration stamps had different values and items for sale had point values based on availability.

To finance the war, our parents were urged to buy war bonds and we were pushed to buy ten-cent savings stamps at school. We never discussed the war, or anything else, at home, but my brother, Jack, and I watched exciting war newsreels at the Saturday morning movies. I remember the newsreel that first introduced us to the new "Jeep" vehicle for combat duty. And, of course, there were lots of war movies in which our troops, usually led by actor, John Wayne, bravely fought the bad Nazis and nasty "Japs"; I still watch the reruns of these old propaganda movies every chance I get.

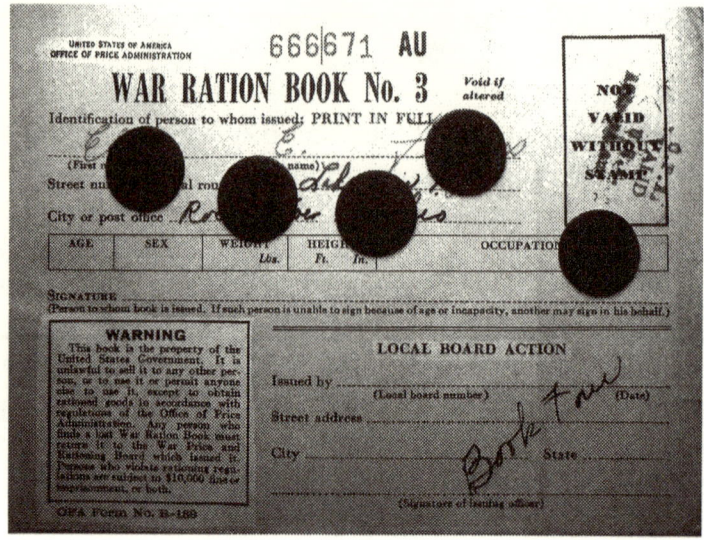

A WORLD WAR II RATION BOOK AND TOKENS.
These books contained stamps that had to be traded to buy meat, sugar and other scarce items. The black dots in this photograph are red and blue ration tokens.

There was newly invented margarine to replace butter (you had to mix the yellow color into it using a powder because the dairy industry would not let them make yellow margarine to compete with butter). We had a "victory garden" in our little tiny back yard to grow our own vegetables. We saved and turned in grease, tinfoil (from chewing gum and cigarettes), toothpaste tubes, old aluminum pots and pans, paper, razor blades, lipstick cases, rags, string and newspapers for the war effort and to earn extra ration stamps. We had neighborhood "scrap drives" to salvage aluminum, steel pipe, rubber and other strategic materials. Posters urged us to "Throw your scrap into the fight," and "Save your cans to pass the ammunition."

I earned a Boy Scout medal for helping to collect 1,000 pounds of newspapers. Sugar was scarce because it was needed for gunpowder, dynamite, and other chemicals. Copper pennies were replaced with steel and zinc so that the copper could be used for battlefield telephone wires. Cooking fat was reused to make glycerin for ammunition. Silk was needed for parachutes, which made silk stockings very scarce for the ladies. So, "recycling" is really nothing new to anyone who lived through World War II. Beef was strictly rationed, but pork was more plentiful. As a result, Hormel Foods invented canned "Spam" (meaning spiced ham) made of a pork blend (nobody really wanted to know what parts of the pig they put into the blend!). Millions of pounds of Spam were fed to American, European and Russian troops and helped to win the war. We ate lots of Spam on the home front too, and I still like it to this day.

During the war, we had a black cat that we named "Black-Out." It was named after our wartime drills during which we had to cover all our windows and all outdoor lights in the whole city were turned off as practice in case enemy bombers should raid the city some night.

Production of metal toys was halted and replaced by wooden and paper toys. Many toys during the war had a war theme and incorporated wartime propaganda against the enemy Germans, Italians and Japanese, who were collectively known as the "axis" powers. All the other good countries of the world fighting against them, including the United States, were known as the "allies."

The favorite wartime toy of my brother, Jack, and me was known as the "Junior Secret Bombardier Set." This included a small, wooden bomber plane with U.S. Air Force markings that had a wooden bomb in a little compartment that could be released by pushing a button on the plane. In addition, there was a hole in the tail of the plane and by looking through this hole into a tiny mirror inside you could look straight down below the plane to aim the bomb. The set also included a cloth target sheet to put on the floor that had pictures of enemy ships, tanks and planes to drop bombs on. Jack and I would walk along the floor sighting into the bombsight of the plane and then drop the bomb on the targets. We spent endless hours doing this during the war and this is the one wartime toy we really remember.

Our apartment at 216 South Meramec Street in Clayton was small and modest. It only had one bedroom, which Jack and I slept in. Mom and Dad slept on a "Murphy In-A-Door Bed" in the living room, which tilted up vertically and swiveled around into a closet when not in use. Jack and I were always fascinated by the contents of the closet with the Murphy bed, because it held all of Dad's old things from Canada like golf clubs, ice skates and skis. While I remember this apartment as being pretty modest, our neighbor below us was a judge at the nearby Clayton County Courthouse so I guess it wasn't too bad.

Standard legwear for boys at this time was brown, corduroy knickers that ended just below the knees with long socks below. The dress of this time is perfectly displayed in the very funny, classic movie entitled, "A Christmas Story," starring Darren McGavin.

We did a lot of our shopping in Clayton at Gutman's Department Store. Especially memorable was the store's x-ray machine to ensure that new shoes fit correctly. This large machine had three viewing scopes: one scope for the salesman, one for us, and one for our mother to look through. Looking through the scopes, you could see the outline of the new shoe and all of the bones of your foot inside. We always had fun wiggling our toes in the process of determining if the new shoes were the right size. Unfortunately, it was discovered in the 1950s that these machines gave off hazardous levels of radiation and they were banned. I don't know how much radiation we got, but our shoes always fit!

In the winter, we got our heat from a coal furnace in the basement that we had to hand-feed by shovel. Once or twice during the winter, a load of dirty and dusty soft coal was delivered to us from a truck. The coal was shoveled down a steel chute through the coal window on the side of our apartment building into the dark and dirty coal bin in the basement. In the basement was also our large, cast-iron, hand-fed, coal-burning furnace. To make this beast provide heat during the cold St. Louis winters was quite a challenge to Dad, and I did my share when he was not at home.

In order to get steady heat from the furnace it was necessary to shovel in the coal and keep it properly stoked—not too much coal or too little. When the heat started to die down, Dad knew it was time to go down the stairs to the basement and add more coal. But first, he had to "shake the fire down" with a big iron handle on the side of the furnace to let the burnt cinders fall through the grate to a pan below, which we had to empty frequently (cinders came in handy to sprinkle on our sidewalk when it got icy). Then the new coal had to be shoveled in and carefully "banked" around the sides of the burning coals so the fire would not be "smothered." In addition, from time-to-time fused chunks of coal impurities called "clinkers" had to be removed with tongs through the furnace door. All of this created a lot of black, dirty coal dust and soot both in our apartment and in the air throughout the city.

Whenever Mom would wash our hair in the winter, Dad would stoke up the furnace for extra heat and we would sit with our wet hair in front of the hot air vents or "registers"; there were no electric hair dryers in those days! At night, the furnace would have to be carefully banked with coal in order to have enough glowing coals to get the fire going again in the morning. I can recall waking up many a morning to find that the furnace had gone out during the night and the apartment was freezing. This meant that Dad had to light kindling and newspapers to get the fire started again. One time it fell to me to start the furnace. I got impatient at the slow progress so I got the idea of using an electric fan to blow on the fire to make it start more quickly. I guess it was a good idea but I made the mistake of letting the fan touch the furnace and it shorted out with a shower of sparks!

During the summer months, Jack and I moved to our screen porch. An oscillating electric fan was the only "air conditioning" we had during the hot, humid summers. I vividly recall waking up in the middle of the night one time to find our electric fan in flames a few feet from my face; I screamed and Mom and Dad came running into our room, unplugged the fan, and calmed Jack and I down.

On Meramac Avenue, Jack and I, and neighborhood friends, amused ourselves with activities and games such as: baseball in the street (until I hit a home-run through a neighbor's window), hide-and-seek, Monopoly, Dominoes, Chinese Checkers, Pick-Up-Sticks, Tiddly-Winks, erector set projects, shooting marbles, playing with yo-yos, and looking for 4-, 5- and 6-leaf clovers. And, then there were sledding and snowball fights in winter, magic shows, cowboys and Indians, and so forth.

Just to give you an idea about our neighborhood games, here is how we played "hide-and-seek." To start, the kid that was "it" had to close his eyes and count to 100, while the others ran off and hid. Then "it" would go looking for the others, and upon finding one would try to outrun him back to the base. If he did so, then the one who had been found became "it" and all the others were called back to base by yelling "Ollie, Ollie, ox in free," which really meant "all outs can come back free." The things that all of our games had in common were that they were fun, free, and helped us learn to get along with our playmates. At Cub Scouts, I learned how to braid Indian leather belts, tie knots (I can still tie a "sheep shank"), and use a drill and wood-burning set to make gifts for our parents, such as ashtrays and cribbage boards (although we had no idea these boards were used for).

Ice trucks still delivered blocks of ice to many homes, which did not yet have electric refrigerators, and a card in the window indicated how many pounds were needed. To get the right amount of ice, the iceman used an "ice pick" to break up large blocks and trim them. We picked up the chunks of ice left on the street and sucked on them and pressed them on manhole covers to pick up the lettering impressions.

We also entertained ourselves in these years before television by listening to afternoon adventure programs on the radio like "Jack Armstrong, the All-American Boy," "Captain Midnight" and "Buck Rogers." The biggest thrill was sending away breakfast cereal boxtops for "decoder" rings and badges; at the end of a radio adventure a series of letters and numbers was read out and by adjusting the two dials in the decoder it would decipher the message into something like "Buy War Bonds" or "Be Loyal and True." This was very exciting and may have led to my lifetime interest in codes, ciphers and espionage. I especially liked signaling with the siren-like whistle in my Jack Armstrong Egyptian Whistle Ring. The wonderful thing about these radio adventures was that we visualized in our minds everything that happened. We had to pay attention to every detail said during the programs in order to follow the action. We did this every weekday and got many thrills. (I still enjoy listening to those old radio shows when they are broadcast today and even have a collection of them on tapes to listen to on long car trips.)

The big event of the week was the Saturday morning movies for kids at our local Shady Oak Theatre (it is still there today). We walked a mile or so to the theater and paid our dime or quarter admission. We then bought our popcorn as excited kids were running all around the place. The lights dimmed and then they would show the latest episodes of the "movie serials" that continued from week to week. There was Flash Gordon fighting in outer space against Ming the Merciless, Detective Dick Tracy fighting big city crime, Tarzan's jungle adventures, and the adventures of Buck Rogers in space 500 years in the future. A full-length adventure film filled with action followed the movie serials. We got our weekly fix of sword-fights, cowboys and Indians, pirates, and battle scenes. There was a metal shop next to the theater and after the movie we picked up long strips of tin and made swords out of them. Then, we played at sword-fighting all the way home and even inside our apartment (our porch screens suffered some bad sword cuts). Exciting stuff for boys!

Our main rainy day activity in the summer as boys was playing in the water flowing in the street gutters. We put on our swimming suits and made leafy dams to form little lakes. Then we would launch our paddleboats, which we had cut out of thin boards of wood in the crude form of a boat. A wound-up rubber band powered the little paddle wheel in the back.

On a rainy day every now and then, my brother, Jack, and I made lead soldiers with our lead-casting set. I still have that old casting set in the attic and I have such fond memories of making shiny soldiers with it that I wrote the following article that was published in the January 1997 issue of *Good Old Days* magazine:

> *As a boy in St. Louis during World War II, it was always a thrill to see shiny new lead soldiers come out of the black sooty molds of my home casting set. On a rainy day, my younger brother, Jack, and I would go down the spiral stairs into our dark basement to a storage locker near our hand-fed coal furnace. After a brief search, we found the box labeled, "Lead Casting Set by Home Foundry Mfg., Chicago, Illinois," and took it upstairs to our apartment.*
>
> *The first thing I did was plug in the small electric ladle, because it took a long time to get hot enough to melt the lead. Jack gathered up our broken and bent lead soldiers, casualties of previous war games. We looked over our selection of molds and chose "The Marines Have Landed."*
>
> *The electric ladle was now smoking, so we dropped in the broken soldiers and watched the arms, legs and torsos slowly dissolve into liquid gray lead. We added some lead pipe, begged from a local plumbing shop, to fill up the pot. As things melted, I skimmed off the smoking scum of dirt and paint.*
>
> *Jack screwed wooden handles on the two halves of the aluminum mold, and then we held them low in the flame of the little alcohol lamp from our Gilbert chemistry set to coat them with black soot.*

holes on to the deposits below. When we came for Sunday lunch, Uncle Bob would shoot a couple of squirrels and Aunt Hattie would fry them up as the main dish; they tasted much like chicken but just a little gamier.

Also memorable from those days were our trips to Tulsa, Oklahoma, to visit our Aunt Peg and Uncle Marian Halsey. Uncle Marian was an independent oil drilling operator or "wildcatter." Although I didn't know what a wildcatter was, I was very interested in the pile of oil-soaked rock cores that Uncle Marian had in one corner of his back yard from wells he had drilled. We also liked to shoot arrows on a nearby archery range in a field that is now the Utica Square Shopping Center. The Halseys would also take us out east of Tulsa to a private club on a lake that they belonged to called The Ozark Club. We enjoyed swimming and fishing in the lake, but one time a Cottonmouth Moccasin snake came swimming by and gave me quite a scare!

We had a lot of superstitions that we paid attention to. While walking on a sidewalk we would try "not to step on a crack because it would break your mother's back." Black cats that ran across in front of you were bad luck, as was walking under a ladder. We carried a rabbit's foot on a little chain for good luck; four-leaf clovers, which we were good at finding, were also considered good luck and carried in our wallets. One superstition that I still follow is saying "bread and butter" anytime I am walking with someone and we walk around something like a pole or tree on different sides. To this day, if I tell someone about something that is going well, I "knock on wood" so as not to bring bad luck.

In 1943, I finished at Meramec School, which only went through sixth grade, and went off to the Maryland School several miles away for seventh and eighth grades. In those "good old days," there was no worry about school violence; the biggest problems for teachers were students chewing gum and running in the halls! After school, we kids went home and said, "See you for dinner, Mom," and then disappeared to play games until dark without a worry in the world.

In the summer of 1945, when I was almost 14 years old, I traveled alone by train to Toronto, Canada, to once again visit my Canadian aunts, uncles and cousins. This 750-mile train trip involved changing trains in Detroit and was quite an experience. I took in a movie between trains in Detroit and almost missed my connection. After reaching Toronto, my relatives and I spent a week or so on a small island in the Muskoka Lakes, north of Toronto. The relative I was most impressed with at the lake was an older cousin, also named Fred Kelly (Fred George Kelly), who at age 22 was working his way through college drawing comic books. I remember his adventure character was named "Rick O'Shay" (after a bullet ricochet). Cousin Fred amused himself that summer by drawing cartoons on the walls of the little island's outdoor toilet.

Due to some kind of ill feelings between my father and his Canadian relatives, which I have never fully understood (possibly because we were rather well off during the early years of The Depression while they had a rough time of it), I did not see or talk to any of my Canadian relatives from 1945 until my visit to Canada in 1998, 43 years later! [But, fortunately, I wrote my Aunt Gretchen Ballantyne, Dad's sister, in Galt, Ontario, in 1970 and 1972 and asked her about Dad's family history. Aunt Gretchen wrote me two long, detailed letters, which provided the basic knowledge about my father's interesting family tree, which I have built on over the years.]

Just as I was ready to start junior high school in the fall of 1945, World War II was finally over. Germany had surrendered in May 1945, and then Japan surrendered in September 1945, a month after we had dropped the first atomic bombs in history on their cities of Hiroshima and Nagasaki. So, I was 14 years old at the beginning of "The Nuclear Age!"

3 ✳

Morse Code and Adventurous Dreams

In the fall of 1945, I traded in my corduroy knickers and long socks for my first long pants and got ready to start high school. For one year (1945-46), I attended Wydown Junior High School. This involved long but fun daily rides in a rickety old wooden streetcar (which we nicknamed the "Toonerville Trolley" from a cartoon strip popular at the time) through wooded, affluent neighborhoods to the school. At Wydown, I became interested in photography and joined the Photography Club where I learned and enjoyed photographic developing, printing and enlarging; as a result, photography is a hobby that I have followed all my life.

In the fall of 1946, I started senior high school at Clayton High School. I rode my bicycle several miles to school every day. Our school was an old fashioned building and, of course, had no air conditioning. In the warm months, the large windows were opened for ventilation; this gave an opportunity for rascally students to climb out of ground floor windows during classes when the teachers weren't looking.

Following the end of World War II, the market was flooded with cheap "army surplus" clothes and equipment. I loved to look through the dark and junky surplus stores (one shopkeeper jokingly told us to "keep our hands in our pockets and whistle," so we would not steal anything!). During the cold winters of my high school years, I wore a surplus, Type B-3, sheepskin-lined aviator's jacket with a large collar, brass hardware and a heavy zipper. I cleaned the sheepskin by rubbing cornmeal into the fur.

I remember very little about my classes, teachers and grades during my three years at Clayton High. I think most of my grades were "C's" and I recall having to take a mathematics course over during summer school (who could have guessed I would end up passing integral calculus in college!). I was much more interested in my scientific hobbies and friends than I was in classes and grades.

During my high school years, my mother had to work to supplement Dad's modest paying job at the fuel yard. She constantly worried about money and rumors that our apartment building was going to be torn down to make room for office buildings and we would have to move. Ironically, our old apartment building is still there to this day, although now turned into a couple of offices, while every other house and apartment that was in our neighborhood was torn down years ago to make room for high-rise office buildings! No doubt I picked up on Mom's constant worry, as I became a bit of a worrier myself.

Mother worked for years as a saleslady at a toyshop and then in a store that sold music recordings called The Record Bar in Clayton. I visited these stores often and for awhile I was paid 25 cents to mail packages at the post office. The couple that owned the Record Bar had a little boy who was always running around the store; that little boy grew up to be Kevin Klein, the popular actor-comedian.

Back in 1943, when I was about thirteen, I had started working during summers at a large public swimming pool in nearby Shaw Park in Clayton. I began as a "basket boy," checking people's clothes in wire baskets while they swam. Over subsequent summers during high school, I moved up to "maintenance man," which was mostly janitorial-type duties, and then to "water engineer" in charge of chlorinating the pool water and cleaning the three huge filter tanks in the basement of the pool house. As water engineer it was my job to test water samples from the pool every hour to check the chlorine level and the "pH" or acidity level. I then adjusted the chlorine pumps, which used cylinders of very dangerous chlorine gas. If I put too much chlorine in the water the complaints came into the manager about the water burning people's eyes and bleaching their swimming suits. If I let the sun burn off the chlorine and get too low then the bacteria count would go up and the manager and I both got into trouble with the City Health Inspector. And, to clean the filters I had to "back wash" water through them in the reverse direction, which resulted in my wading around in water three feet deep in hip boots. I also threw a burlap bag full of blue, copper sulfate crystals into the pool every night to help clear up the water for the next day.

After school during my later years in high school, I was an usher at the Shady Oak Cinema in Clayton. This small movie house showed "art" films such as: "Hamlet" with Lawrence Olivier (I was to see him live on stage in London at the "Old Vic" theatre some 20 years later); "The Red Shoes," about the London ballet; and the movie of the opera "La Traviata." The Shady Oak is still showing movies to this day.

I enjoyed all of my part-time jobs during high school. My fellow workers and I always had lots of fun on the job. At the pool, we snapped wet towels at each other and invented all sorts of games and rituals. One time I rigged up a battery and spark coil from a Ford car and wired up a stack of wire baskets to shock the customers; that got me into trouble with the manager! My part-time jobs also provided me with enough money for my hobbies and entertainment. Money was really not a worry for me personally, although I was aware that Mom and Dad were concerned about making ends meet.

After our family had been forced by adverse circumstances during the Depression to move into our apartment in Clayton, Mom and Dad both had to work and they had little social life. All of their former society friends, except for one very kind couple (the Cadles), either abandoned them or were avoided by Mom and Dad because of embarrassment over their poorer circumstances. And, most of the other kids I went to school with at Clayton High School were from comfortable, middle or upper-class families with large homes, while our family lived literally "across the tracks" in a small apartment building amidst old Victorian houses. Our family was short on money, and had no car. I guess it all gave me somewhat of an inferiority complex, which I have wrestled with most of my life.

I avoided most social life at high school and never dated or went to dances. My pals felt the same way I did about social life even though they came from comfortable middle-class homes, so I guess it was not just a matter of money and status. Our parents somehow convinced us social life was something to be avoided. Many parents of my friends were constantly yelling at each other, even when I was visiting. Although my pals and I didn't participate much in high school social activities, we still had fun. We went to movies and drive-in restaurants, cruised around, an occasionally got into minor mischief such as staging fake kidnappings on street corners and moving people's garden furniture to other houses. Once in a while we would venture into downtown St. Louis and sneak into the Grand Burlesque Theatre with its program of alternating striptease artists and rude comedians. After the burlesque show, we would watch the strippers in nearby bars (we could not buy drinks at our age, but the understanding bartenders would give us empty beer bottles to hold).

I still keep in touch with some of that "old gang of mine." Jim Knudstad became an archaeologist and is still going out on "digs" in the Middle East and North Africa. He lives in a little village in Cornwall, England, and, together with his wife, Rosa, also an archaeologist, is restoring a 500-year-old cottage using original granite bricks and buying period stone windows and doors. He has even designed arrow slits over the front door and admits he is still living out his boyhood fantasies.

My oldest friend in the world (since 1940), Merle Wolff, worked his entire career for a railroad in St. Louis and is now retired; we still enjoy reminiscing about the old days in Clayton. Another old friend is a retired dentist. They both married and have families. One of my junior high school friends was very "artsy" and became a theater set designer on Broadway in New York, but he died many years ago (according to his obituary, he was survived by his friend, George!).

My big passion during high school was tinkering with radio equipment. I started by building crystal radio sets that used no electricity but required long aerials and a ground wire clamped to a water pipe. I had to move the tip of a little steel "cat's whisker" around on the surface of a piece of galena lead crystal (a primitive forerunner of the transistor) until I could hear a station; the signal was weak and only worked with earphones. I mostly heard classical music from a nearby religious radio station but it was very exciting at the time! Then I started building one- and two-tube radios from kits, using large, glass radio tubes, and suddenly I was hearing all the local stations!

I gradually advanced in radio until I had my own licensed amateur (ham) radio station. My station call letters were WØROI, and I operated using really neat, surplus, military radio equipment sold cheaply after the war (WWII). With my 100 watts of power coming from a surplus ARC-5 aircraft transmitter, I talked to hams all over the States by microphone. But, I usually preferred tapping out Morse Code messages because using this mode I could reach ham stations all over the world from my little station on the back porch of our Clayton apartment. My surplus BC-348 aircraft short-wave receiver was a beautiful piece of equipment. Even today, I turn on my short wave radio every now and then, and copy some Morse Code just for fun; Morse Code is like riding a bicycle, you somehow never forget how to do it.

I was also an avid short-wave listener during my high school radio days, and regularly listened to the British Broadcasting Corporation (BBC) in London and other world stations. The station that really fascinated me was Radio Australia, which always signed on with the wild cry of the Cookabura Bird. I got so wrapped up in Australia that I subscribed to the *Sydney Morning Herald* newspaper from Sydney, Australia, and sent away to Sydney for a "digger" hat with a brim that snapped up on one side. I bought a large boomerang and got pretty good at throwing it; it would circle far out of sight, re-appear as a thin profile headed right for me, and then pass menacingly right over my head. During this "Australian Phase," I even visited a nearby elementary school and gave a talk on Australia. Little did I suspect that I would actually visit Australia some 30 years later.

While in high school, I bought a 22-foot-long bullwhip from a Boy Scout supply house and became a very early "Indiana Jones." I would walk around with it coiled over my shoulder, and then unsling it and snap it at all sorts of objects like trees and plants, and occasionally friends. With all my strength put into the 22 feet of whip, it would crack like a rifle, which I am sure all the neighbors really appreciated!

I had two basic whip-cracking techniques: one was a forward snap with the right arm, while the other involved twirling the entire whip over my head several times and then reversing my arm direction to give a tremendous crack that you could hear for a city block.

Other activities I engaged in during high school would be considered pretty dangerous today. These included: BB-gun fights (I got hit right next to my eye once), throwing knives, bows and arrows with steel hunting tips, fiery chemistry set experiments, home-made gun powder fired in cannons made from steel pipe, carbide (acetylene) miner's lamps and carbide cannons, stove bolts screwed together with match heads (using Ohio Blue Tip kitchen matches that had big phosphorus heads that would strike on anything, and explode if packed too tightly) in between that exploded when thrown on the sidewalk, pea-shooters, firecrackers, flattening pennies on railroad and streetcar tracks, and other rambunctious behavior.

Guns were items of fun and fascination. The father of my good friend, Jim Knudstad, would take us out to a limestone quarry on the bank of the Missouri River where we would shoot all our guns. I fired away at floating tin cans in the quarry with: my single shot .22-caliber rifle; Dad's .45-caliber Colt automatic from World War I, and a cheap .38-caliber revolver I bought somewhere (the cylinder did not quite line up with the barrel and it sprayed lead off to the side; it was a wonder it did not blow up!). We especially enjoyed firing our guns under a bridge over the Missouri River as the result was loud, successive echoes off of each bridge support. I kept a drawer full of guns and ammunition in our apartment and I don't recall my parents ever paying any attention to it and I certainly never worried about it.

I was involved in a drive-by shooting once, however. One Halloween night in my senior year I was cruising around with my gang in a 1947 Chevy. We had a high-powered water-pump and drove around squirting people on the street with a terrific stream of water.

During high school, I was a Boy Scout but not a very enthusiastic one. I mostly remember the weekend camping trips, when it seemed to either pour rain and flood our tents or snow overnight and freeze our drinking water. One trip included a hike through fields of green, bell peppers, which we picked and ate along the way; I am still fond of bell peppers and always remember that field when I eat one. One summer our troop

ME AND MY GUNS.
A highschool era photograph of me with my .22 caliber rifle and my
father's World War I, .45 caliber Colt automatic in its holster.

went by train to a summer scout camp at Irondale, Missouri. One rainy day we chased a young counselor around the campsite and when I caught him by the ankle he stabbed me in the arm with a dagger. I still have the scar on my left arm. I also vividly remember our scoutmaster parading us around in the nude; whenever I hear about alleged sexual abuse by Boy Scout leaders, I can believe it!

Another thing I really enjoyed in high school was the year I spent with the Junior Achievement program. We formed up into companies and the company I joined decided to make and sell imitation vanilla extract, a flavoring used in cooking. We designed a business plan and then bought our raw materials; the main ingredients were vanilla flavoring and propylene glycol, as I remember it. We made our imitation extract, bottled it, put our own labels on the bottles, and then sold the final product door-to-door. I still think of that project every time I see a bottle of vanilla.

I did not go out for sports in high school. However, in my junior year the football coach said I had the build for football and asked me to join the squad. I took up the game with vigor and became a guard or tackle. I greatly enjoyed charging through the line and throwing the opposition quarterback to the ground, often ripping off his shirt in the process. Unfortunately, I seriously pulled a muscle in my groin just a few weeks into the season and that ended my very brief football career.

Jack and I attended Sunday school at Central Presbyterian Church on Hanley Road in Clayton during high school. Mother came with us to church once in a while, but Dad was an Episcopalian and would not attend with us. We did go to an Episcopal Church with Dad once or twice, and he was very religious and quite puritanical around the house. There was never any bad language, nor did Jack and I run around the house without clothes on.

Mom and Dad were both very gentle and soft-spoken. I cannot recall hearing a single argument or disagreement between them. They also never complained even though their quality of life had fallen considerably since their halcyon days in the early 1930s. We never discussed family matters like money, religion, sex, or any other emotional issue, but few families spoke about these things in those days. There was also very little show of affection around the house, which I am sure influenced my character. I blame most of this situation on Dad, who was very quiet and conservative, while Mother was a very warm, outgoing and friendly person. Again, Dad was some twenty years older than Mom, which was no doubt a factor too.

One day as I was shopping at Gutman's Department Store, the owner, Mr. Gutman, stopped me and said in a serious tone, "Fred, you know you have to look after your mother." This was a heavy message for a teenager to bear and I have remembered it all my life. I guess I was getting many formative messages at this time and I have pretty well followed these guidelines all my life.

During our family days in Clayton, Mother always served up tasty meals although she was working too. She served things like: chuck roast cooked in a pressure cooker and carved at the table by Dad on Sundays; meatballs with long-grained rice sticking out; pork chops in tomato sauce or cream of chicken soup; fried chicken; tuna and rice with mushroom soup sauce; meatloaf; ham, creamed cabbage, Kraft Dinner; corn-meal mush fried for breakfast with maple syrup on top; waffles with bacon inside; French toast; Kool-Aid; ice-box cookies; banana cake (which we always had on birthdays); Jell-O; and tapioca pudding. We never complained about Mother's cooking, and I still like all the foods she served. I do not recall ever eating in a restaurant as a family.

Dad had a good British-type sense of humor and loved practical jokes. One April Fool's Day after he had gone to work we found he had tied all the dining room chairs together under the table. And, we always had a number of magic tricks and practical joke items around the house. Any friend that came over for dinner was in for trouble because we had all kinds of tricks at the dinner table, such as: a sugar spoon with no bottom, a dribble glass with little holes in the design that trickled water down the chin, and chewing gum that snapped your finger when you pulled it out of the pack. We put pellets on Mom and Dad's lighted cigarettes that filled the room with "snow," and loaded their cigarettes with small explosive charges.

Mom and Dad were both heavy smokers and the sickening mess their cigarettes made around the house resulted in my never taking up smoking. Mother later died of emphysema from all her years of smoking, and smoking could have contributed to the terrible strokes that ended Dad's life.

As 1949 arrived and my high school graduation approached, I was beginning to wonder what the future held in store for me. Just to give you an idea of what was going on in America at that time: the first television sets were just coming on the market with 10- and 12-inch, black-and-white screens and manual control knobs; 78 RPM records were the norm; the Dow Jones average was 179; a new car cost $1,400; Studebaker, Nash and Hudson cars were still around; a small new house cost $7,500; all airplanes had propellers; schools were segregated by race; and drive-in movies were popular. In high school, boys took workshop, welding and mechanical drawing classes, while girls studied cooking, home economics and sewing; these were practical courses that could lead to a job.

I graduated from Clayton High School in May 1949 and went back to my summer job as a "water engineer" at the Shaw Park public swimming pool in Clayton.

All through high school my best friend, Jim Knudstad, and I gloried in the knights of old and dreamt of adventures in the mysterious Middle East. Rather than concentrating on our classes, Jim and I invented a

**MY HIGH SCHOOL GRADUATION
PHOTOGRAPH (MAY 1949)**

mythical foreign kingdom called "Swartzmania"; the name could have
been influenced by the large Jewish presence at Clayton High. I was
the King of Swartzmania (my insignia was "K of S") and Jim was my
loyal cohort. We sent each other elaborate parchment scrolls in old-
English script and burnt the edges to make them look old and crusty.

Our main sources of inspiration for foreign adventure were the books
of Richard Halliburton and Lawrence of Arabia. Richard Halliburton
was a romantic adventurer of the 1920s and 1930s. He wrote a number
of books with wonderful titles such as: *The Glorious Adventure, The
Royal Road to Romance, The Flying Carpet,* and *New Worlds to Conquer.*
These books tell the stories of his worldly adventures, which included:
swimming the canals of Venice in the nude, hunting tigers in Bengal,
riding an elephant over the Alps (as Hannibal's army did in 219 BC),
tracing Cortez's arrival in Mexico in 1519 AD, and flying a biplane
across the Sahara Desert to Timbuktu. In 1939, he disappeared during
a typhoon in the Pacific Ocean while single-handedly sailing a Chinese
junk home to California. A very heroic end for such a dashing figure!

Our other hero, T.E. Lawrence, known as "Lawrence of Arabia,"
kept us transfixed with his books, such as: *The Seven Pillars of Wisdom,
Revolt in the Desert,* and *Oriental Assembly.* These books record his

courageous Middle Eastern adventures while carrying out his Arab revolt against the Turks and Germans during World War I.

To paraphrase Richard Halliburton, Jim and I read books that taught us to "love ages that had past and places far away." As you will see later, Jim and I both made it to the Middle East and all around the world, but we did it in quite different ways.

4 ✳

Genesis of a Geologist

At the beginning of my senior year in high school there had been no thought or discussion in my family about my going to college as there was no money available for this purpose. Then, in the fall of 1948, I had been visited by my Uncle Marian Halsey, an independent oil operator or "wildcatter" who lived with my Aunt Peg in Tulsa, Oklahoma, "The Oil Capital of the World." Uncle Marian and Aunt Peg (my mother's sister) Halsey had always helped Mother take care of Jack and me with moral and financial support (and lots of advice!).

On this visit, Uncle Marian had dramatically covered a table with oil-soaked rock specimens; oil well electric logs and rock sample logs; and oilfield maps. Then, with a confident air, he carefully explained to me what they were. Following his presentation, he asked me what was to be one of the most critical questions in my entire life. He asked if I would like to come to Tulsa to live with he and Aunt Peg, and go to the University of Tulsa to become a petroleum geologist. Thinking the geological data looked very interesting, and having no other plans for college, I immediately said, "Yes."

And so in the fall of 1949, after graduating from high school, I packed up my clothes and took the train to Tulsa, Oklahoma. I then found myself living in the Halsey's guestroom, and what a change for me! I went from a shared bedroom (with my brother, Jack) in our little Clayton apartment to an immaculate old-fashioned, pink, feminine guestroom with frilly lace and old family photographs. This was to be my home for the next four and one-half years while I attended the University of Tulsa. (The room changed very little over those years; no college posters or pin-ups in Aunt Peg's immaculate guestroom.) Not only did I have a completely new bedroom, I also inherited a completely new and different set of parents for the next four and a half years. The Halseys had not acted as parents since they had taken in my mother some 20 years earlier, so this was going to be a new experience for them, and me.

My Aunt Peg, Katherine (Milleson) Halsey, was my mother's second oldest sister. She was born in Nebo, Kentucky, in 1897, so she was 52 years old when I arrived in Tulsa. Aunt Peg was a very proper and proud southern lady just like her mother, my Grandmother Lela, and there was never any doubt that she ruled the household. She ruled with a quiet and fair hand, but also firmly. Aunt Peg's main interests were her marriage, running the house, her close friends, her relatives and her church. She was a good southern cook and managed to put 20 pounds of weight on me over my college years with her. Every delicious meal ended with her telling me to eat everything up because "it won't keep." My primary chore for her was taking out the garbage, a job that I still have to this day.

Uncle Marian, or Marian Albert Halsey, was born in Independence, Kansas, in 1895, and so was 54 years old when I arrived. His father owned the one prominent dry goods store in Independence and his grandfather had been a stagecoach driver. Marian joined the U.S. Army in 1917 at age 22 years and served as a supply sergeant in France during World War I. After the war, Marian became an oilman in Kansas and Oklahoma oilfields. He was a pioneer independent oil promoter or "wildcatter" (defined by an unknown author as "a man alone with nerve to follow a hunch to drill oilwells in unproved territory out where the wildcats yowl"). He was from the breed of tough men who founded the oil industry in America in the 1920s and 30s. They had survived after years of rough industry fighting over oil leases and competition in marketing oil. Marian had worked with and was a good friend of

William G. "Bill" Skelly, founder of the major Skelly Oil Company in Tulsa. Skelly was a tough old "S.O.B." oilman, too. I met him once at his ranch north of Tulsa and he took one look at me and said, "Don't just stand there, get a bucket of paint and paint that barn," and I am sure he was only half joking.

Marian and Aunt Peg had lived in various "oil boom towns" and he had drilled wells all over Kansas, Oklahoma, Arkansas and Louisiana in the 1930s and 1940s, which found little oil but earned him "dry hole promotion money" from neighboring oil companies which gained knowledge from his dry holes. Finally, he discovered the small East Lisbon oilfield in northern Louisiana and also established producing oil and gas interests in the nearby Athens oilfield in partnership with Skelly Oil Company. His oil income provided a comfortable, but not lavish, living for the Halseys in their nice little two-bedroom house at 2208 East 23rd Street in Tulsa (my new home). They belonged to the prominent Petroleum Club and were active in The Central Presbyterian Church in Tulsa.

As you will see, Uncle Marian devoted his four and a half years with me to toughening me up and building up my self-confidence. He left my college studies up to me and never asked about them, but at home he supervised me with a caring but very firm hand. He was a short man and also thought his first name of Marian was rather feminine (he often spelled it "Marion"); as a result of his stature and oilfield experiences, he acted very tough with other men. But, when he got home it was a different story. My Aunt Peg had him firmly under her control at home, and every time he got worked up over anything Aunt Peg would say, "Now Marian!" and that was the end of that.

In the fall of 1949, I enrolled at The University of Tulsa as a freshman to earn a Bachelor of Science Degree in geology from the School of Arts and Sciences. The Halseys paid for the college fees; I think the tuition was only something like $500, even though it was a private college.

I had very little interest in college social life, but Aunt Peg decided I should join a fraternity. I went through the fall "rush" parties and was accepted to pledge with Alpha Tau Omega, one of the top fraternities, which formed in the South just after the Civil War. This was fine with Aunt Peg and so I pledged Alpha Tau Omega, or "ATO." I completed

MY AUNT PEG AND UNCLE MARIAN HALSEY.
My "new parents" in Tulsa, Oklahoma, starting in the fall of 1949.
Uncle Marian was an independent oil operator or "wildcatter." Aunt
Peg was my mother's sister. They had no children of their own.

pledging and was initiated into the fraternity. I got the "full treatment" on initiation, including being paddled and left stranded with a couple of other pledges out in the country somewhere in the middle of the night. (The next school year, ATO did away with "hazing" during initiations because a couple of pledges died after being hit by a car while tied up in the middle of a country road in some other state.)

The Tulsa University campus was some five miles away from the Halsey house, so Uncle Marian brought a car up from his oilfield operations in Louisiana for me to drive. It was a well-used 1939 Ford sedan with stick shift on the floor and mud-tires; nothing fancy, but it got me through my college years. But, before I could use the car, Uncle Marian had to teach me how to drive. He took me out in the country around Tulsa and gave me driving lessons.

My study habits from high school were not very good and I had to struggle through my college classes. Mother always blamed my learning problems on the "progressive" school system back in Clayton, Missouri, which included very little homework. I had to re-start mathematics at T.U. with intermediate algebra.

My social life was not very exciting, but that was the way I wanted it. I attended ATO fraternity functions and dances, but with little enthusiasm. I did no dating, except for "arranged" dates for a couple of fraternity dances. I studied with my fraternity brothers and the fraternity "files" were quite helpful on exams. However, I mostly enjoyed doing things with my fellow geology students.

I continued with my hobby of amateur radio. Uncle Marian built me a really nice, custom designed, plywood desk for my radio equipment. My radio station was set-up in the attic of the Halsey home, surrounded with Uncle Marian's memorabilia, including: his World War I equipment (gas mask, shell casings, binoculars, etc.); oil drilling bits and rock cores; a Civil War rifle and sword; snow-shoes and Indian paraphernalia. The attic had an Indian theme as Uncle Marian knew several Indian artists (such as Woody Big Bow) and admired their work; sometimes when these artists were short of money they "pawned" their paintings with Uncle Marian and they hung in our attic until they were "redeemed." From that attic I contacted amateur radio stations all over the world, mostly with Morse Code that could reach farther on my relatively low-power (100-watt) transmitter.

Meanwhile, Uncle Marian continued "toughening me up." I preferred to stay up late studying and doing ham radio, and then was slow at getting

MY AMATEUR RADIO STATION IN TULSA.
Using surplus military radio equipment from World War II, I was able to contact "ham" radio stations all over the world by voice and Morse Code.

up in the morning (a pattern I still adhere to today). So, Uncle Marian would come bursting into my bedroom early every morning telling me to get up and get going as time was wasting!

In January 1950, the Halseys took me on a memorable trip to northern Louisiana. We drove from Tulsa down south to Shreveport and checked into rooms at the old Washington-Youree Hotel. The purpose of the trip was for Uncle Marian to visit various oil operations in the area in which he had financial interests. Uncle Marian hired a black driver, named Andrew, whom he usually used on these visits, and we drove north out of Shreveport. We visited a gasoline plant in Rhodessa, Louisiana, and then arrived at the Athens Oilfield near Homer, Louisiana. The Athens Field was operated by Skelly Oil Company, but Uncle Marian had helped discover it and owned working interests in many of the producing wells and new wells then being drilled. Needless to say, we got royal treatment from all the Skelly managers at the Athens Field Office. We toured the field facilities and watched wells being drilled as Uncle Marian asked many questions about how much oil was being produced. This was very fascinating to me and gave me my first real look at the "oil patch."

Back in Shreveport, the Halseys took me to dinner at the nightclub in our hotel. After a nice dinner, the floorshow started with various singers and dancers. Then, I was amazed to see a man come out cracking a 20-foot-long bullwhip, much like the one that I had had in high school. The man went through his act of snapping cigarettes in half while in someone's mouth, snapping flowers in half from a vase, and other stunts, while all the time making loud cracks with his whip. Then, he paused and asked if anyone in the audience would like to come up and try cracking his whip. With a little encouragement from Uncle Marian, I got up and volunteered to try his bullwhip. He handed me his whip and then warned everyone to stay well clear of me while I tried to use it. I took the long whip, and just as I had done many times, I swirled it around over my head several times and then reversed direction with my arm and the whip made a terrific crack. This was not at all what the artist had in mind! I guess he expected volunteers to end up with the whip wrapped around their necks, which would give the audience a big laugh. He quickly took the whip away from me and thanked me for coming forward. The next morning, Uncle Marian got me up early to spend another day in the oilfields. Over breakfast he revealed to me that, after the nightclub show, he had asked the bullwhip artist to join us for the day's adventures, but unfortunately he could not make it. Uncle Marian had a way of surprising people with ideas like this.

When Christmas arrived that year, I took the train home to St. Louis to spend the holidays with Mom, Dad and Jack. And, then in May my first year of college was over. I went home to Clayton and once more went to work for the summer at the Shaw Park swimming pool, this time as a night watchman. I alternated nightly with another night watchman and so spent all night alone at the pool every other night. I carried a big eight-cell flashlight to be used as a light and as a sort of weapon, but there was not much real threat of anyone breaking into a swimming pool at night because all money was removed every evening and there was very little crime in Clayton anyway. During the long nights around the dark and deserted pool, the bright stars got me interested in astronomy and I built my first astronomical telescope.

My only real excitement as a night-watchman that summer was walking out to the pool early one Sunday morning to find a fully clothed man edging out on the high diving board. He seemed drunk and said he was going to commit suicide! I talked him into coming down and going home, and then called the pool manager. Some minutes later, the manager and police arrived and gently scolded me for not holding the guy at the pool until they arrived.

In June 1950, communist North Korea invaded democratic South Korea to begin the Korean War. The United Nations declared this invasion to be an "illegal act" and immediately ordered the forces of 16 countries to fight with South Korea. President Truman ordered American troops to join the fight and activated the Selective Service System to "draft" civilians into the army. I, however, was automatically exempted from military service because I was attending college.

In the fall of 1950, I moved back into my room in the Halsey home in Tulsa for my second year at The University of Tulsa. I was happy to discover that the Halseys had just bought their first television set as network programming, such as "The Milton Merle Show," had really started catching on in American homes in the past year or two. Their TV set was a Zenith cabinet model with a 12-inch circular black-and-white screen. My favorite viewing was the Saturday night wrestling matches with such colorful characters as Gorgeous George, who strutted around in his silk robe, and Mr. Moto, who performed a Japanese tea ceremony in his corner of the ring.

As the school year began, several of my fellow geology students and I decided to switch over to the School of Engineering to earn a degree in geological engineering known as a Bachelor of Geology, or "BG." Only two other schools in the country granted the BG, namely Colorado School of Mines and Rolla School of Mines in Missouri. Most geologists earned a Bachelor of Science (BS) in geology from an arts and sciences school. The switch to the Engineering School meant I was now carrying 18 course hours a semester instead of 13, and many of the new courses were tough ones, such as advanced mathematics (calculus), qualitative and quantitative chemistry, and engineering physics. Fortunately, I had little interest in social life so I could now devote even more time to study, especially as passing grades did not come easily for me.

I suspect that the main motivation for my switch to engineering was that I wanted to look and act like an engineer. This meant that I now always wore khaki pants and engineer's boots, had a slide-rule hanging at my side in its scabbard, and belonged to the Engineers Club. As an engineer I also got to play at sneering at students in other school departments and ignored most social events on campus. My grades slowly improved, but with much effort required.

One activity that I particularly enjoyed was our geology field trips to interesting rock outcrops all over Oklahoma. A particularly fascinating place for me was the Eagle-Pitcher lead and zinc mine at Cardin, Oklahoma, northeast of Tulsa near the Kansas and Missouri borders. Walking around the mines and dumps of waste rock, we could easily find beautiful mineral specimens, such as cubes of galena lead and long crystals of brown sphalerite zinc ore. Over the next couple of years my friends and I visited this area several times.

In addition to college activities and Uncle Marian's tutorials on the oil business, I continued operating my ham radio station. I also participated in fraternity functions like building floats for parades, inter-fraternity football matches on the campus lawn, and studying

FIELD TRIP TO A LEAD AND ZINC MINE.
My fellow geology students and I don hard hats and have our rock
hammers ready for action at a mine in northeastern Oklahoma.

THE TULSA ASTRONOMY SOCIETY'S 12-1/2-INCH TELESCOPE.
That's me posing with the Society's homemade, 12-1/2-inch diameter
telescope, which was quite large by amateur standards.

with my brothers who were also in the Engineering School (the fraternity's test files were also helpful). And, the bright stars over the Halsey house inspired me to expand my interest in astronomy. I built a rather nice, 3-inch, reflecting astronomical telescope and mounted it on a war surplus tripod. I enjoyed viewing Jupiter's moons, Saturn's rings, nebulae and other sights. Then, I read a newspaper article about The Tulsa Astronomy Society and immediately joined up. The Society was a nice group of people and they had built a large 12-inch telescope using steel oilfield casing; this really opened up my viewing of smaller planets and other celestial phenomena. I also met a fellow college student named James A. "Jim" Westphal, who became a close friend during my college years.

Jim Westphal was older and had worked for an oil geophysical crew in Mexico before returning to college at Tulsa University to complete his degree in physics. He was big into astronomy and I discovered he was living alone in an apartment near the Halsey home. His apartment was more of a physics laboratory than living quarters, and he was always building something. I spent a lot of time over at his place and we also did lots of celestial viewing with the Society's big telescope located in a nearby neighborhood. Jim was a near-genius with a photographic memory; when I would ask him if he was ready for an exam the next morning he would say, "Oh, is there an exam?" and then he would make an "A." [Jim went on in life to become a renowned astronomer and helped design the Hubble telescope.]

As my second year of college came to a close in the spring of 1951, Uncle Marian said that he would like for me to spend the summer down in the Athens Oilfield in northern Louisiana. He said this would accomplish two things at once. First, my main job would be to look after operations in the field in which he had a financial interest and report back to him daily. Secondly, it would be good experience for me in petroleum geology and oilfield operations.

When the college year ended, Uncle Marian and Aunt Peg drove me back down to Athens, Louisiana. Uncle Marian provided me with a car and settled me into a small, furnished apartment over the garage next to a nice little house that was also the post office for Athens, Louisiana. My new landladies, Gussie and Lena Mae Harris, were elderly spinsters and very gracious southern ladies; they also ran the post office. Uncle Marian then took me a few miles down the road to the Athens Oilfield Field Office of Skelly Oil Company. He again introduced me to Graham Hankins, the manager, and outlined what he wanted me to do for the summer. I am sure Hankins was not thrilled about having me hanging around all summer, but Uncle Marian was a working interest owner in many wells and helped him pay his bills. Hankins told me he had appointed his field geologist, Bill Short, to look after me and train me.

Aunt Peg and Uncle Marian returned to Tulsa and I got started on my new summer job. I visited the Athens Field Office every day and used the daily well reports to prepare a written report to be mailed to Uncle Marian every day. In addition, I worked with Bill Short as he described rock cuttings and cores from the wells, correlated electric logs, drew contour maps, and did all the other things that petroleum geologists do. I observed all "drill stem tests" run on the newly drilled wells; these tests consisted of opening the valves on the wells up to flow oil or gas out into a pit where it was set on fire for disposal. The flow of oil or gas was measured with instruments, samples were taken, and charts from the bottom of the hole were read to find out the bottom hole pressures. This was all quite exciting for me as many tests were very high pressure gas flows which made loud screaming noises, and the reduction of gas pressure caused all the flow lines to ice up like a refrigerator coil.

In the middle of the summer, Uncle Marian asked me to go into Shreveport to meet with Mr. Austin Stewart, owner of Stewart Drilling Company and an old friend and partner of Uncle Marian's. In his office, Mr. Stewart explained that his children were in a summer camp in the Hill Country north of San Antonio, Texas, and it was time for he and his wife to pick them up and bring them home. He said he would like to have his car in San Antonio to drive up to the camp and spend a couple of days in the hills before flying back to Shreveport. He offered to pay me $10.00 a day to drive his Cadillac from Shreveport over to San Antonio and then back again. This sounded exciting so I readily agreed. I felt very grand as I drove away from Shreveport in Mr. Stewart's big Cadillac Fleetwood headed for San Antonio, Texas. I drove west to the Texas border and then through East Texas to Austin and then on to San Antonio; it was a pleasant but uneventful trip. I delivered the

**A STEAM DRILLING RIG IN THE ATHENS OIL
FIELD IN NORTHERN LOUISIANA.
I visited this and several other rigs every day to compile my
daily report to my Uncle Marian Halsey in Tulsa.**

car to the Stewarts at the San Antonio airport and they dropped me off at The Hotel Gunter where I stayed until they returned with the car a couple of days later. They flew home and left me with the car to drive back.

I had asked Mr. Stewart if I could take a couple of extra days on my way back for my first visit to the Gulf of Mexico at Corpus Christi. Mr. Stewart agreed and so I drove south from San Antonio to Corpus Christi, where I checked into the rather primitive Hotel White Plaza, which had a nice view of the Gulf; but no air-conditioning, of course. I was fascinated by the shipping and took a ferry over to Aransas Pass, where all ships entered and left the port. I watched the ships with my 30-power Wollensak telescope, which I always carried with me in those days. I was standing next to a small lighthouse and was delighted when a Coast Guard officer called down and invited me to come up and watch the ships from his control room. This is where I learned the expression "she has a bone in her mouth" to describe a ship under speed with a big foamy bow wave in front.

I returned to Shreveport and handed the car keys back to Mr. Stewart. He asked me to return the unused emergency money he had given me for the trip. I had to sheepishly admit that I had spent more on my personal travels than I had planned and that there was very little of his extra money left. Mr. Stewart frowned and looked a bit disappointed but said he would call it even. I then left his office and drove back to Athens to resume my duties for Uncle Marian there.

Towards the end of the summer, a severe heat wave hit the area with daily high temperatures of 104 to 105 degrees. With no air conditioning, it was hard to get comfortable in my apartment during my afternoon break. I solved this problem by spending a couple of hours every afternoon in a bathtub full of coolish water reading paperback books. I then re-emerged at sundown and went back out to the Athens Field Office to see what was going on. Some evenings I drove up to Homer, Louisiana, the county seat of Claiborne Parish and a nice little town built around a city square with an old courthouse and a Confederate Army war memorial. I ate at a nice restaurant there and often took in a movie too.

I also enjoyed driving to the Driskill Mountain forest fire-watching tower some 25 miles southeast of Athens. I would climb up a couple of hundred feet of steps to the top of the deserted tower to watch the sunset over the pine forest. This tower is located on the highest point in the state of Louisiana (elevation 535 feet above sea level).

The summer of 1951 came to an end and I returned to the Halsey home in Tulsa for my third year of college at The University of Tulsa.

5 ✴

Final College Years

In the fall of 1951 I started my Junior Year at the University of Tulsa. My Engineering School course load for the year of 18 credit hours a semester included advanced geology courses, differential and integral calculus, and engineering physics. I really don't know how I made it through all that calculus and physics because it did not come easily for me. In any case, my results for the year were about equal "B" and "C" grades. I continued to participate in ATO fraternity affairs like building a float for Homecoming and studying hard with brother engineering students.

Tulsa University's football team was on a roll and was chosen to go to the Gator Bowl. As a result, our paleontology (fossils) professor drew very complicated drawings of plant and animal fossils on the blackboard and made us copy them by hand because "there is no money for copying paper since the football team needs new uniforms!" This situation rather soured me on college athletics.

Uncle Marian continued giving me his own personal course in confidence building and the oil industry. On weekends we would drive out into the country to visit oilfields and old friends of his, like Bill Hoge up in Claremore, Oklahoma, who had been a good friend of the late humorist, Will Rogers. Hoge formerly lived in Oolagah, Oklahoma, until the state built Grand Lake and flooded the town out. He was the town barber and wrote a newspaper column called "Oolagah Oozings"; his picture was published in a *National Geographic Magazine* article on the area.

The down side of Uncle Marian's teachings was that he was rather intolerant when it came to black people and Democrats. He really disliked Democrats, especially President Franklin Roosevelt and his "New Deal." One time the two of us were looking through an old almanac in the attic and came across a full-paged photo of Roosevelt; Uncle Marian let out a curse and ripped the page out of the book and threw it away. Unfortunately, Uncle Marian passed on some of his intolerance to me and that became a problem for me for a long time.

The spring of 1952 was interesting as the Geology Department teachers took us on geology field trips to fascinating rock outcrops in the Arbuckle-Ouachita Mountains of southern Oklahoma and along the Illinois River in northeastern Oklahoma. Then, just before school let out in the spring of 1952, Professor Murray, the Dean of Geology, passed the word that Gulf Oil Company wanted to hire four senior students for the summer to help with a petroleum geology survey of the White River Valley of northern Arkansas using aerial photographs. I jumped at the chance and told Dr. Murray to give Gulf my name.

I was contacted by the Shreveport Office of Gulf Oil Company office in Shreveport, Louisiana, and told I had been selected for their summer program, along with three of my fellow students. All four of us were told to report to their Shreveport office as soon as school was out. When the school year ended, the four of us attended a meeting in Gulf's Shreveport Office during which the geological survey program was outlined. We were told we would be living in Fort Smith, Arkansas, so we could field check the work done from the aerial photographs. Two Gulf geologists were in charge of the project and they drove us to Fort Smith, Arkansas, where they had rented a house at 515 North Greenwood as our office. They then showed us the boarding rooms they had rented for us in an old Victorian house owned by an elderly lady.

Our job that summer was to match up overlapping aerial photographs and then interpret the geology on them to look for oil prospects. We studied the photos through large desktop stereoscopes and then traced the results from the photos on to maps using light tables. There was no air conditioning in those days and it got very hot working over the tracing tables that used a large number of high wattage, incandescent light bulbs. We worked stripped to waist and sweated profusely.

STEREOSCOPES AND AERIAL PHOTOGRAPHS.
We used these tools to view the geology in three dimensions
in order to map oil prospects in northern Arkansas for
Gulf Oil during our summer job in 1952.

Fort Smith is right on the Arkansas-Oklahoma border. At the time, Arkansas had alcohol prohibition and was "bone dry," while Oklahoma also had prohibition but did allow low alcohol (3.2 percent) beer. So, after a full day sweating over our light table, we would walk over the bridge at Fort Smith into Oklahoma where they had a series of "beer joints." Believe or not, this is where, at age 21, I first learned to drink alcohol!

Our work that summer also involved car trips into the White River Valley north and east of Fort Smith to field check the geology we were seeing on the photographs. In the summer heat, we drove and hiked around the countryside and visited remote farms, often wondering if we were going to be shot as trespassers. When we met people we told them we were gold prospectors so they would not get excited about the possibility of big oil finds! We also visited Arkansas towns like Fayetteville, Eureka Springs, and Harrison during our trips. Towards the end of the summer, the two Gulf geologists took us on a very interesting geology field trip to Hot Springs, Arkansas, where there are some very unusual mineral ore deposits. Here barite ore is mined for use in oilwell drilling fluids and there is also a large deposit of magnetite. Magnetite is a highly magnetic iron ore and airplane compasses were said to go crazy when planes flew over the area.

The summer job with Gulf paid well and the money went towards books and other miscellaneous college costs; the Halseys paid my tuition and provided room and board as for previous years. The summer job with Gulf was also very good experience. Meanwhile, my brother, Jack, had graduated from Clayton High School in the spring of 1951. And, with financial help from Aunt Peg and Uncle Marian Halsey, he attended college at Washington University in St. Louis. In the fall of 1952, he transferred to Westminster College in Fulton, Missouri. Westminster is well known as the place where British Prime Minister Winston Churchill gave his famous speech in 1946 in which he referred to the Soviet Union and its Eastern European satellites as being apart from the world "behind an iron curtain." At Westminster College, Jack majored in economics, joined the Sigma Chi Fraternity, played baseball and joined the U.S. Army Reserve Officers Training Corps. (ROTC). This college activity deferred him from being drafted to fight in the Korean War.

MY BROTHER, JACK.
In 1952, while a student at Westminster College, Fulton, Missouri.

The Korean War ended with an armistice in July 1953 but the "draft" continued. I began to wonder how my exemption from military service would play out after I graduated from college. Soon, however, my summer job was over and it was time for me to go back to Tulsa for my fourth year of college. My fourth year at the University of Tulsa consisted of advanced geology and geography courses adding up to 18 or 19 credit hours a semester. Fortunately, I now had all my engineering courses in mathematics, chemistry and physics behind me. The most interesting course that year was two semesters of economic geography. The instructor was an older woman who was in charge of the geography department. She made her course very interesting and I can still remember many of the things she said about the ways that geology and geography affect the economy of various regions of the world. The most interesting study concerned the Paris Basin in France. Various sandstones, limestones and shales outcrop on the surface in a circular pattern around the Paris Basin. The type of rock in any one region forms a unique soil and these soils produce all of the different kinds of famous wines found in the basin just outside Paris.

In addition to my studies, I continued with my fraternity, ham radio, and astronomy. I also made numerous geology field trips in the Tulsa area, both school sponsored and with fellow geology students, who had now become close friends. One very interesting project that school year was a course in geological field surveying using an alidade instrument and a plane-table mounted on a tripod. After learning the techniques in class, I was assigned with two good friends, including Jim Westphal (my astronomy buddy), to survey and map the roads, houses, creeks, oilwells and topography in a one square mile area north of Tulsa. This square mile was officially designated as Section 12, Township 20 North, Range 12 East, Tulsa County, Oklahoma.

First of all, Jim Westphal got the brilliant idea of buying an official county aerial photograph of our square mile. This photo essentially drew all the oilwells, houses, roads, creeks and ponds in the area for us, so all we had to do was survey in a number of check points to make sure we correctly plotted everything. This left us with the primary job of surveying a network of points all over the section and calculating the elevations at these points. We then contoured these points as topography on our map. The job of hand drafting the final map was left to me as I had a good hand for drawing and lettering. I still have this map in my possession today, and it looks pretty good, if I do say so. We only got a "B" grade for this work, but the map looks like "A" grade material to me even now!

OUR FIELD SURVEYING CREW.
My fellow students sitting on my 1939 Ford car as we
surveyed a square mile north of Tulsa, Oklahoma.

My course results for this fourth year increased to several "A" grades and the rest mostly "B's", with a couple of "C's." This improvement had a lot to do with not having any more engineering courses, but I have always felt that I have been a "late bloomer" at many things. I had learned how to study more effectively as the years of college went by.

A man lived up the street from us in Tulsa that Uncle Marian knew. His name was Roland Stanfield and he was a close relative of the Williams family that owned the huge, worldwide Williams Brothers Pipeline Company in Tulsa. Roland Stanfield was a senior officer with Williams Brothers, a company known all over the world for its major pipeline projects. His young son often came over to see me at the Halsey house to look through my telescope at the stars and examine my rock and mineral collection. I got the son started on his own rock and mineral collection, and gave him my entire collection when I graduated from Tulsa University.

Roland Stanfield had heard from my Uncle Marian that I had just completed a course in topographic mapping and he contacted me just as my fourth year of college neared its end. He said Williams Brothers desperately needed a topographer for a pipeline project they were just starting in the Tioga-Beaver Lodge oil and gas field located at Tioga, North Dakota. He said that there were not any plane-table topographers left in the oil industry as topographic maps were now made from aerial photographs with field checks.

Stanfield asked me if I would spend the summer in Tioga making a topographic map of the Tioga-Beaver Lodge oil and gas field. He said Williams Brothers would then use my map to lay out a natural gas pipeline to a gas gathering plant for Signal Oil and Gas Company. The gas had a lot of clear gasoline-type liquids in it and so the pipeline had to be laid downhill all the way so the liquids would drain into the plant where they could be collected. He offered me a very generous salary of about $500 a month for the summer.

I agreed to do the job for Williams Brothers. Right after school ended, I took Greyhound buses all the way north to Tioga, North Dakota, and reported to the Williams Brothers field office there. I was assigned to a senior engineer, Ben Hasha, for my topographic project. Ben showed me aerial photographs of the Tioga-Beaver Lodge Field area that I could use for my topographic mapping. The whole area was covered with wheat fields with scattered roads leading into oil and gas wells. I then noticed that every farmer's field had a very noticeable pile of rocks in it. I found out that a glacier had covered the area tens of thousands of years ago, and as it melted away it dropped rocks and boulders all over the countryside. Over the decades, the farmers had cleared these rocks and used to them to make fences, but after every rain more rocks would emerge, which they now threw onto rockpiles. I decided these rockpiles would make excellent places to locate my survey instruments because they were so clearly visible on the aerial photographs; this way I only had to measure elevations and not worry about locating the positions of the stations on my maps.

Ben hired two high school boys on summer vacation to help me with my surveys and it turned out one was of Norwegian background and the other German. As a result, I learned to give them orders in both of these languages, which often resulted in much laughter. We started to work with a known elevation at an official survey point and then carried this elevation into a field using my plane-table and alidade survey set and a stadia rod held by the boys. Once in a field at a point located on the photographs, usually an easily identified rockpile, the boys carried their rod all over the nearby fields and I surveyed in elevation points, which I then contoured into topographic maps.

I used a small hand level to roughly sight on the terrain and plan where to locate survey points. I decided it would be nice to have a leather holster so I could wear the level on my belt. I went to the local shoemaker, an elderly Norwegian with a thick accent. I sketched out what I wanted him to make for me and he took one look at the drawing and said, "That's a bunch of damn monkey business!" But, he went on to make the holster and did an excellent job of it; it is as good today as the day he made it. It rained continuously that summer and the mosquitoes were terrible. The little yellow devils would bite our faces and hands and right through our trousers. I located some mosquito nets at a war surplus store and we wore them all the time. As we worked in the wheat fields we also kept a mouthful of what the locals called "Russian peanuts," or sunflower seeds, which were grown locally. We would move the seeds from one cheek, crack them and eat the seed, and then spit out the shells. At the local movie theater a sign read, "no chewing gum or Russian peanuts allowed!"

We worked every day that summer from sunrise to sunset because the pipeline project was on a tight schedule and the pipe laying was approaching the areas where they needed my maps to plot the downhill gravity flow. As the summer went by, Ben could see I would not get the maps completed before I had to go back to college, so he hired three men to make up another survey crew for me.

I trained the new crew and then had two crews to supervise, and twice as many survey points to contour every evening. This was also my first experience at supervising men older than I was, and it was an education. I found out that they complained to each other about how I moved them around and the many hours I made them work. I also kept an eye on the crews from a couple of low hills that I used to help sketch in the topographic contours. I would pull out my trusty Wollensak

ONE OF MY FIELD SURVEY CREWS AT WORK IN TIOGA, NORTH DAKOTA IN 1953.

Shooting in elevations stations in a wheat field near Tioga, North Dakota, using a plane-table and alidade.

FIELD CONTOURING ON THE STEPS OF A GAS WELL IN TIOGA, NORTH DAKOTA.

After my crews had surveyed in elevation stations, I would view the terrain and draw in elevation contours on my maps. My maps were used to plan a route for a gas pipeline.

30-power telescope and see what the crews were up to. Sometimes the crews were lying in the wheat taking naps instead of working, so I would drive over and tell them to get back to work, as I had seen Uncle Marian do with his oilfield workers many times. Needless to say, this did not further endear me with the men.

Tioga was virtually an oil "boomtown" with muddy streets and lots of bars. The small local populace was mostly of Norwegian origin and when I went into a store they would be speaking Norwegian. I lived in the newly constructed Clyde's Boarding House over a laundromat. I shared a room with another man and every one in the place was an oilfield or pipeline worker. On Saturday nights everyone got drunk and threw beer cans up and down the long hallway leading to the rooms. None of this bothered me much because I worked almost 14 hours every day of the week and was dead tired when I went to bed and fell asleep immediately.

Ben tried to convince me to go to work for Williams Brothers as a surveyor after I graduated from college. To this end, Ben assigned me to the pipe laying crews for a week for experience. I followed the surveyor around as he staked out the pipeline route and right-of-way. Then I spent a day on the pipe bending machine crew that bent the pipe to fit the curves in the route. Finally, I put on hip-high rubber boots and held the surveyor's survey rod in a creek crossing with water and mud almost waist deep. I decided the pay was excellent on a pipeline crew but the work and hours were real killers!

As the summer ended I had most of the topographic mapping finished but there was so much data coming in from the two crews that I could not find time to do all the topographic contouring. So, Ben hired a college student to learn contouring from me so he could complete the maps after I had returned to college. I trained him in the art of topographic contouring. When my summer work was finished, I took a bus back to Tulsa with a stop in the Black Hills of South Dakota to see Mount Rushmore and visit the Deadwood Saloon where Buffalo Bill Cody was shot in the back. I arrived back in Tulsa ready to start my final semester at Tulsa University.

I had now completed four years of college but I needed one more semester to take various elective courses to make up a shortage of hours caused by my switch from the College of Liberal Arts to the Engineering School back in 1950. One elective I chose was a course in botany because I thought it would be good for a geologist to know about vegetation and plants. This turned out to be rather humorous because the rest of my class were football players looking for an easy credit and housewives who wanted to know about growing flowers around their houses. They were all completely boggled by the various Latin terms for plant parts that they had to learn, but these terms were very easy for me after four years of complex geology, chemistry and physics. When the teacher announced test results, she would say there was one score of 99 percent (mine), which would be thrown out, and scores of 75 to 85 would be an "A," and then down the scale. Needless to say, the other students were a bit uncomfortable with having me around and openly complained about me to the teacher!

I worked after school for a petroleum geological and geophysical consulting firm, called Manhart, Millison and Beebe, which was good experience and brought in some extra money for college expenses and ham radio equipment. I also won first prize in the first engineering school technical writing contest with a paper I wrote entitled, "Petroleum Exploration with Radio Waves." This paper combined my interest in radio with my petroleum geology. I delivered my paper before a large audience and received a check for $50 as my prize.

Meanwhile, the Ohio Oil Company (later renamed the Marathon Oil Company) had offered me a job and I finally decided to go to work for them after graduation. My decision was based largely on their promise that they would have assignments in North Africa coming up in the near future. I completed all my courses in January 1954 and as a mid-term graduate went to work for Ohio Oil in Ardmore, Oklahoma. I was officially awarded my diploma for a Degree of Bachelor of Geology from the School of Engineering at the University of Tulsa on the evening of May 31, 1954. For some reason, Mom and Dad did not attend my graduation, but Aunt Peg and Uncle Marian were there. And, so, the college phase of my life was officially ended. However, as I was already at work in Ardmore, my transition from college to the work place was very smooth.

SIX MUNCE UGO I CUTNT EVN SPEL GEEOLIJIST AN NOW I ARE ONE ...

6 ✳

Arbuckle Oil

Following graduation from the University of Tulsa in January 1954, I reported for work at the Tulsa, Oklahoma, division office of the Ohio Oil Company. Here they told me that I was being assigned to their regional exploration office for southern Oklahoma located in Ardmore, Oklahoma, near the Texas border. Once again I reiterated my desire to go to the Middle East, and I was told to be patient and gain experience in Ardmore. They said they were working on obtaining oil concessions in Libya, North Africa, and expected to be sending field geologists over there in a couple of years.

My move from the Halsey's home in Tulsa to Ardmore, Oklahoma, was really my first move out on my own. Knowing this, Aunt Peg and Uncle Marian Halsey set out to establish me in Ardmore ready to go to work. First they helped me buy and finance a used Plymouth automobile in very good condition. This large, sturdy, purple car was quite an improvement over the 1939 Ford with mud tires that I had been driving all through college. (I was to find out later, however, that every time I drove through deep water after a heavy rain, the water would splash up under the hood and kill the engine. I had to sit and wait until the water evaporated off the hot block.)

I packed my few belongings into my Plymouth, and the Halseys followed me from Tulsa to Ardmore in their car. The Halseys knew just what to do next because in their younger days they had spent many years traveling to small towns following oil booms. They read the classified ads of the local paper and found me a room to rent in a large, old Victorian house owned by an elderly lady. I ended up in a huge, old-fashioned room with minimal furniture, but I did have my own door to the outside.

Now that I had a room and a car, I was ready to go to work for Ohio Oil's office in Ardmore as a petroleum geologist at a starting salary of $402 a month. Based on all of my summer oil field experience while in college, I was started at a salary equivalent to having been with the company for a year. This was a very good starting salary as a small house could be bought for under $20,000 in those days. I reported for work at the Ohio Oil offices in the Bowman Building on Stanley Street near the courthouse in Ardmore. The offices were in a long, narrow, two-story office building. The district geologist had an office in the front with a secretary's room attached. A coffee room followed this and then there was a long hall with little offices off each side, one of which was assigned to me.

The district geologist sat me down and told me what all the managers told fresh geologists right out of college. He told me that for five years I wouldn't be worth a darn to the Ohio Oil Company as I would be training and learning on the job. He said that after the five years I would be ready to do some valuable work for the company. He then assigned me to a senior geologist who would take me under his wing. The senior geologist took me into his office and for some weeks we went over the oil and gas maps of southern Oklahoma. He showed me where the Ohio Oil Company had wells drilling and what rock formations they expected to find oil in. Then he put me to work "running samples," which involved examining rock cuttings out of previously drilled wells and logging rock descriptions on a long strip of paper marked off in feet. The idea behind this was to train my eye so that I would be able to recognize these formations when I went out "sitting on wells." I ran samples like this for some time; in fact, running rock samples under the microscope is a basic job that many petroleum geologists perform throughout their careers.

In a few months, I was assigned a county in southern Oklahoma and given the base maps with all previously drilled wells plotted on them. My responsibility was to: keep track of all wells drilling in my county, run well samples, examine electric survey logs on wells, mark rock formation "tops," and write formation sub-sea elevations under each well symbol on the maps. Then, I was to contour the maps on various geologic horizons to try to discover undetected anticlines which might contain oil, or what we called "prospects," that Ohio Oil could drill. The area that I was assigned was a part of southern Oklahoma north of Ardmore and west of Winnewood and Paul's Valley. This was an oil and gas producing area, which was called "The Golden Trend." Oil had already been discovered in many fields in ancient sandstones some two miles under the ground.

AT MY DESK IN ARDMORE, OKLAHOMA.
My first job after graduation as a petroleum geologist
with The Ohio Oil Company in 1954.

A few months later, Ohio Oil started drilling three wells in my area on locations that had previously been staked out by other Ohio Oil geologists. I was assigned the duty of "sitting" on these three simultaneously drilling wells. It seemed that I would not have to wait five years to start doing valuable work for Ohio Oil! My job was to visit the wellsites and describe, or "log," the rock cuttings or samples as the wells drilled and check them for traces of oil and gas. If I saw traces of oil, I would call my boss in Ardmore and then order the rig crew to cut a solid rock "core." I would then examine the core for oil or gas and send samples to a laboratory so their porosity and permeability could be measured to see whether they would be good oil producers.

To get to my drilling wells, I drove my company car north over the Arbuckle Mountains to Winnewood, Oklahoma. Every time I crossed the Arbuckle Mountains, I was enthralled by the fact that the rock layers I was driving over were the same ones that a few miles north contained oil at depths of one to three miles below the ground. Arbuckle geology was truly fascinating!

For my well site work, I camped in a small, dilapidated motel in Winnewood, which is about 20 miles north of Ardmore. My motel room was so small that the toilet was located inside the shower stall. While my three wells were drilling, I had round-the-clock duties visiting one well after the other. I examined rock cuttings and gave orders to the drillers that if they had a "drilling break," or sudden fast drilling that indicated soft sandstone with oil potential, they should stop drilling and call me to come and check the samples. Following a drilling break, I would wait until the rock cuttings were circulated up to the surface by the drilling mud, which could take half an hour or more. I then checked the cuttings through my microscope for stains of brown oil that would dissolve in a carbontetrachloride solution and glow yellow under an ultraviolet lamp. If these signs were present, I knew that the drill bit had encountered an oil-bearing sandstone.

If I had a definite oil show, I would call my boss in Ardmore and get his approval to cut a core. Then I would order the driller to go into the hole with a "core barrel," instead of a drill bit, to cut a 60-foot column of solid rock, with a diameter of three and a half inches, for detailed examination. When a core came out of the well, it was all hot and steamy and covered with dark brown drilling mud. Using my rock hammer, I would break the core into convenient sized pieces about six to eight inches long. I would then examine the pieces with a hand lens, smell them for oil, and taste them for oil and/or salt water. Smelling and tasting rocks for oil is the time-honored practice of petroleum geologists. I would also look to see if there was any gas bubbling out of the core that might indicate natural gas in the formation either combined with oil or as gas alone. I then boxed up the core to send to a storage facility in Ardmore.

I well remember that one night I was asleep in my motel room while a rig crew was cutting a core with instructions to call me when they started pulling it out of the hole. I woke up in the middle of the night, sat up in bed, and had a premonition that the core was coming out. I got dressed and drove out to the well, and sure enough, the core was on its way out. The driller had finished cutting the core and had forgotten to send anybody to let me know. Drillers rather resented being told what to do by young geologists and often did things, either consciously or subconsciously, to show the "rock hounds" who was really in charge! Geologists that got too "uppity" often ended up in the mud pit!

If one of my cores had oil in it, then a petroleum engineer would come out from Ardmore and we would run what was called a "drill stem test." During this test, we would go into the hole, set a rubber plug just above the oil zone, and then open up a special valve at the bottom of the drill pipe to let gas and fluids from the formation flow into the empty drill pipe. Then we would wait to see if oil, gas, or both would either come up to the surface or fill up the drill pipe to a certain height. This gave us a measure of the reservoir's content, pressure and flow capacity.

Running a drill stem test was always exciting. We often waited all night until the sun came up so we could turn off all the electric lights to eliminate any fire hazard. Then at dawn we "opened the tool" and watched a little rubber hose attached to the drill pipe that was immersed in a bucket of water. If the hose started to bubble, we knew something was flowing into the drill pipe. If the bubbling got very vigorous, then we knew we had a lot of something. We then started smelling the hose to see if natural gas had reached the surface. After an hour or so, we closed the valve at the bottom of the drill pipe by rotating the pipe, came out of the hole, and waited to see what we had recovered. Most of the time in Oklahoma there was not enough pressure for the oil, gas, or salt water to flow to the surface, so we would unscrew the drill pipe and find maybe a few hundred or even a few thousand feet of either oil or salt water or both. Oil was very good news and exciting, but salt water was bad news because it meant that there was very little oil in the formation.

So, I went from well to well running the samples. In those days there were no fancy geological laboratories at the well sites as these came along some years later,. All I had was a car, a microscope, some carbontetrachloride solution, some hydrochloric acid and an ultraviolet light box. I would run the samples under my microscope out on the fender of the car. At night I would run the samples by holding the microscope in front of the headlights.

After drilling several thousand feet of new hole, we would stop and an electric logging truck with a huge spool of electric cable would come to our wellsite. We would then run electrical survey logs in the hole, which would indicate the types of rock formations that we had just drilled and give an indication of whether or not they contained oil or gas in commercial quantities. This was the moment of truth for me because the electric logs had the effect of checking my work watching the rock samples and indicated whether or not I had missed any showings of oil. Fortunately, I never missed an oil zone, but making sure this did not happen was a heavy responsibility for me.

For several months, I was driving around from drilling well to drilling well around the clock. I pulled cores, ran samples, ran drill stem tests, and supervised electric logs with very little sleep. I usually ate at a little diner on the outside of Winnewood, and I followed the old oil-field adage that, "If you can't sleep, then eat." So the less I slept, the more I ate. Another oil-field tradition that I frequently practiced was sleeping in my car when I had to wait for something to happen in the middle of the night or at daybreak; this was referred to as "sleeping in the Chevrolet Hotel."

As 1954 unfolded, I settled into a routine of petroleum mapping, sitting on wells, and taking geologic field trips around the area "to check the outcrops," that is to examine the rock exposures. All of this involved the complex geology of southern Oklahoma, which is some of the most fascinating in the world. We found oil too, which made it all the more exciting.

On the personal side, I enjoyed doing things with other oil company bachelors. I soon found out that several of them were living in a large, old apartment house at 319 North Washington Street in Ardmore. One of my friends, also a geologist, asked me if I wanted to move in with him and share expenses. I thought this would be a lot more interesting than living with an elderly woman in an old Victorian house. And so, I moved into the apartment building, which housed mostly single men and women, and this made life a lot more interesting. There was no air conditioning in those days so our windows were open in the summer. My bedroom was right next to a traffic light on a highway through town, so all night long I heard trucks downshifting and starting and stopping at the light, but it didn't seem to bother me.

I bought a used 10-inch, black-and-white television set and I think it was the only set in the apartment house. Every night we would get a small crowd in our little apartment to watch the "Tonight Show" with Steve Allen until it went off at midnight. I settled into the habit of staying up until midnight or later and then dragging into the office half asleep in the morning. (I followed this pattern of sleepy, late mornings throughout much of my 38-year career with Marathon Oil, except for the years in London and Pakistan, where the offices opened at 8:30 or 9:00 a.m. and we lived very close to the office. While I started a little late in the mornings, I fully made up for it by working evenings and weekends.)

My other social activities in Ardmore included square dancing, YMCA volleyball, and dating a couple of local girls. I also took up snorkeling and spearfishing in the murky waters of nearby Lake Murray. One day, I chartered a small Piper Cub airplane to fly me around over the nearby Arbuckle Mountains. I got some excellent photographs of the spectacular geology and scenery.

The big topic of conversation and gossip in early 1954 was the televised McCarthy congressional hearings on alleged Communist influence in the government, military, and Hollywood. Television was a relatively new and exciting thing at this time and these hearings really put television "on the map." After all my college years with my archconservative Uncle Marian, I supported McCarthy's efforts to get rid of all those "Commies," and argued about it with some of my fellow office workers.

During my final year at the University of Tulsa I had won first prize in an engineering school technical writing contest with my paper about exploring for oil with radio equipment entitled, "Petroleum Exploration with Radio Waves." After graduation, I was asked by *The Petroleum Engineer* magazine to prepare my paper for publication in their magazine, so I would drive out to Lake Murray Park and work on my paper. My six-page article entitled "Petroleum Exploration with Radio Waves" was published in the June 1955 issue of *The Petroleum Engineer,* and I was paid the princely sum of $100 for my efforts. It is ironic that the one and only technical paper I ever had published during my career was based on a paper I wrote in college.

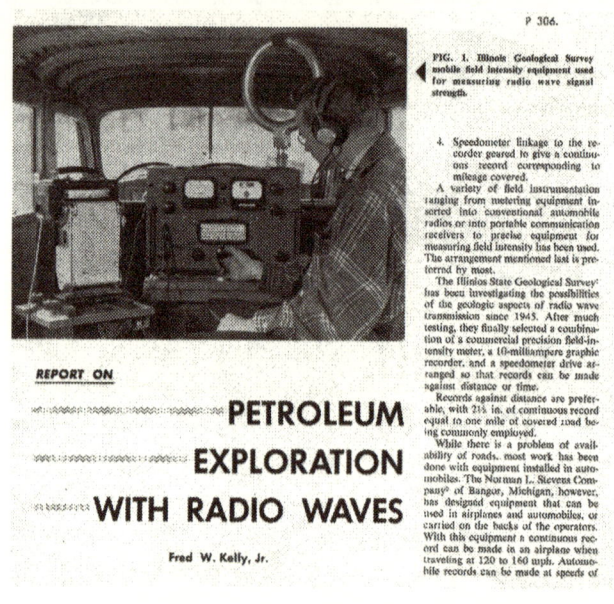

TITLE PAGE OF MY PUBLISHED TECHNICAL PAPER.
A six-page article with many maps and diagrams that was published in the June 1955 issue of *The Petroleum Engineer* magazine.

Some months after my paper was published, the Petroleum Exploration Society in Shreveport, Louisiana invited me to present my paper to them. And, at about this time, I decided to carry out some experiments on my own time. I had a radio signal-strength meter wired into the car radio in my Plymouth and planned a geological survey in the Arbuckle Mountains, where the geology was well known. I drove around on paved roads and recorded radio signal strengths from local broadcast stations. I then plotted these readings as profiles on graph paper. It appeared that I could see a change in signal strength over known faults, although I didn't actually map any potentially oil-bearing geological structures. It was an interesting experiment, nevertheless.

By this time, I was getting a reputation within my company for tinkering around with strange equipment to map geology. As a result, the company sent a geologist from its Bakersfield, California, office to Ardmore to show me his technique for mapping geological faults using a magnetometer. A magnetometer is an instrument that measures the magnetic force from igneous rocks buried deep in the ground. It is normally used to take broad magnetic readings to give a regional view of the geology, nothing like mapping anything as detailed as a single fault. However, this geologist showed me that by taking magnetometer readings 100-feet at a time across an area he could in fact map a fault zone. He theorized that magnetic materials were deposited in the fault zones by circulating ground waters. Following a short course of instruction from this geologist, our office contracted a magnetometer crew and put me in charge of testing it in the Arbuckle Mountains to see if I could map known faults. If I could do that, then we could use the magnetometer to survey unknown areas for faults that could trap oil.

My crew and I started mapping along an area of known geology up in the Arbuckle Mountains. Reading the magnetometer consisted of taking the instrument out of its case, mounting it on a tripod, and leveling it very, very carefully. Then we would very carefully lower a balance on to a quartz prism. Then, by reading the instrument, we could observe very, very minute changes in the earth's magnetism that reflected the type of rocks in the ground. We mapped along a road for a couple of weeks without any significant change in readings. Suddenly, the readings went completely wild and way off scale. We stopped and read the instrument every hour, but the readings continued to be very erratic. The magnetometer crew chief called his office to ask what could be happening since they had never seen anything like it before. His office studied the problem and then informed him that we must be experiencing a magnetic storm in the earth's magnetic field. They said to keep reading the instrument every few hours until the storm stopped.

The erratic readings went on for several days. Then one of the crew noticed that one of the balance weights on the magnetometer was loose; apparently the lock nut had worked loose. We suddenly realized that we weren't dealing with a magnetic storm; it was just that every time we took the instrument out of the case, the weight got jiggled to a different position and gave us wild readings. We tightened up the weight, recalibrated the instrument, and didn't have any trouble for the rest of our survey. We began to pick up some interesting readings, but after the so-called "magnetic storm" fiasco my company lost enthusiasm for the project. The project was abandoned with inconclusive results, but it had been a valuable experience for me.

MAGNETOMETER SURVEY IN THE ARBUCKLE MOUNTAINS OF SOUTHERN OKLAHOMA (1955).
With this crew I was experimenting with mapping geologic faults using magnetometer readings. That's me on the right. A loose weight in the instrument sabotaged the project.

At this time, my brother, Jack, had been inducted into the U.S. Army and, after finishing basic training, was posted to Erlangen, Germany, for 18 months. While in Erlangen, Jack wrote me that a good, but somewhat eccentric, friend of his had just "crashed" actress Grace Kelly's huge wedding to Prince Rainier in Monaco using an invitation he had forged from a newspaper photograph. Meanwhile, I had been visiting Mom and Dad in St. Louis, and while there I visited occasionally with my very good, high school friend, Jim Knudstad, who was now an archaeologist. Jim and his family were driving a Jaguar sports coupe and a British MG sports car, Model TD. The more I thought about that MG, the more I decided I wanted one.

Sometime in 1955 I drove my Plymouth to Dallas, Texas, located about 100 miles south of Ardmore, to a foreign-car import company and picked out a nice, royal blue 1954 MG, Model TD, that was used but in excellent condition. I traded in my Plymouth for it and financed the rest. The car needed a few minor repairs, and so it was agreed that I would come down and pick it up in about two weeks. During the two weeks back in Ardmore, I began to worry about what I had bought and what condition it was in. Buying this sports car was a big move for me at the time and one of my first independent financial decisions. Two weeks later I drove back to Dallas, got into my MG-TD and started north towards Ardmore just as it got dark. I really didn't know how to drive the car or where the controls were. The first problem I ran into was trying to find the headlight dimmer switch. I looked all over the steering column, on the floor, and on the instrument panel but couldn't find it, so I pulled over to the side of the road and started looking around the driver's footwell. Lo and behold, the dimmer switch was in the upper left-hand corner of the foot well. I had to reach up with my left foot to dim the headlights.

As months went by, I grew to love my little MG. I would take the carburetors apart and clean them and I would set the spark plugs. At that time, there were very few foreign sports cars in southern Oklahoma, so my MG was quite an oddity. People were always stopping to look at it. While it looked powerful, it actually had a rather small 50 horsepower engine; but, as long as I kept changing gears, I could keep it moving pretty fast and it cornered well. On geologic field trips I'd drive my little MG into small towns way out in southeastern Oklahoma. When I would drive down Main Street, all eyes were on my MG and when I parked everyone would gather round to look at it. Alcohol prohibition was in effect throughout Oklahoma in those days with only 3.2-percent beer allowed. While I was not much of drinker, I would occasionally drive my MG to Wichita Falls (known as "Whisky-ta Falls"), Texas, and smuggle a small load of liquor back to Ardmore.

I met another bachelor geologist who had recently returned from military duty in Germany and had brought back a German Porsche sports car. The two of us used to hang out together and double-dated with our sports cars. I had a straight-line muffler put on my MG, and when I stepped on the accelerator, the thing really roared. On February 23, 1956, I appeared on the front page of the local newspaper with one wheel of my MG in a giant pothole in the road; the newspaper was campaigning for pothole repairs and recruited me to help as my small car accentuated the pothole.

**A NEWSPAPER PHOTOGRAPH OF ME AND
MY MG SPORTS CAR IN A POTHOLE.**
The local newspaper in Ardmore, Oklahoma, ran this
photograph in 1956 to illustrate the local pothole problem.
My small car accentuated the pothole.

In August 1955 the company assigned me to be their "client's representative" on a seismograph crew operating in eastern Oklahoma near McAlester. This is an area of surface coal mining, or strip pits. My job was to keep an eye on the crew and also to learn how seismic surveys were carried out. I spent about a month on this crew. In those days, a small drill truck drilled a hole 10-20 feet deep, which was then loaded with dynamite. The dynamite was exploded electrically, and the shock waves from the blast traveled down through the earth, reflected off the rock layers, and were detected back on the surface by a set of very sensitive seismometers some distance away, which recorded the readings as wiggly lines on a photographic film. By measuring the time delays we could calculate the depth to the rock formations. [It's all done without explosives these days and is much more sophisticated and computer controlled.]

While we were "shooting" our seismograph along the roads, the seismograph crew had a lease man who went ahead of us and paid the landowners for permission to survey across their land. One of the landowners told him that we could shoot across his land and also said that we could fish in his strip pit. His strip pit was full of water and stocked with large-mouth bass. I bought my first fly rod and an assortment of fake grasshoppers and other dry flies that floated on the water. At sunset, I cast these flies out on the water, twitched them a little bit, and almost every time a big bass would come up and grab one. It was really great fishing. I've never been much on eating fish, so I gave them to the other members of the crew, who had their families with them living in house trailers and boarding houses. They were quite happy to get my nice, big, large-mouthed bass.

In the meantime, I had been keeping in touch with Merle Wolff, my old grade school and high school buddy in Clayton, Missouri. Back in high school, we had organized a summer bus trip to Miami, Florida, which we both enjoyed. He was now working for a railroad in St. Louis, and we decided it was time for us to take another trip. I recalled stories I had heard while in college from my friend, Jim Westphal, who had worked on a seismograph crew in Yucatan, Mexico. Based on these stories, I organized a trip for Merle and me to Mexico City and the Yucatan Peninsula of Mexico. The details of this adventure to Mexico's "Well of Death" are recorded in the next chapter.

In the fall of 1955, Mom and Dad celebrated their 25th wedding anniversary with a trip to Canada to visit Dad's relatives. I spent Christmas at home that year and heard all about their nice trip.

**MOM AND DAD ON THEIR 25TH WEDDING
ANNIVERSARY TRIP TO CANADA IN 1955.
Dad was 64 years old and Mom was 45 years old at this time.**

In early 1956 Ohio Oil sent out questionnaires to all of their geologists asking them if they wanted to go to Libya to do fieldwork in the Sahara Desert. This was just what I had been waiting years for. I filled out the questionnaire on both sides with all the reasons why I wanted to go to Libya. I then signed the questionnaire in Arabic as a Jordanian student at the University of Tulsa had taught me. My boss said that my questionnaire signed in Arabic created quite a stir in Ohio Oil's headquarters in Findlay, Ohio, and that I would probably be assigned to our Libyan operation in the near future.

At about this same time my college draft deferment that had kept me out of the Korean War expired because I was no longer in college. Even though the Korean War had ended in 1953, the Selective Service, also known as the "Draft Board," was still drafting soldiers and I was notified that I had been reclassified "1-A" and was immediately subject to two years of military service. Fortunately, I had read in a newspaper about a new Army Reserve program that was designed for scientific and professional personnel with "critical skills." This "SPP" program allowed professional personnel to serve six months active duty in the army and then spend seven and a half years in the inactive reserves with no meetings or anything else. Petroleum geologists were listed as eligible for this program. When I appeared before my Draft Board in Tulsa, Oklahoma, they told me I was now "1-A" and would be inducted for two years of active military service. I told them that I wanted to apply under the "SPP" program. The Draft Board had never heard of such a program, so I showed them my newspaper article. They did some quick checking and then admitted, "You're right; there is an 'SPP' Program and you are eligible for it."

On September 4, 1956, I reported to Fort Bliss, Texas, near El Paso, for eight weeks of basic training. The details of my basic training and the remainder of my rather short military career are reported in a subsequent chapter.

7 ✳

Mexico's Well of Death

In early 1955, my old friend, Merle Wolff, and I decided it was time to plan another adventurous trip together. Our high school trips to Meramec Caverns in Missouri, to Chicago, Illinois, and to Miami, Florida, had been very enjoyable. At this time I was working for the Ohio Oil Company in Ardmore, Oklahoma, and Merle was working for the Missouri-Pacific Railroad in St. Louis.

At my suggestion, we decided to fly to Mexico City for our next adventure. We started working out travel arrangements and things to see and do. I also suggested a side trip from Mexico City over to Merida in Yucatan, Mexico, to visit the Mayan ruins at Chichen Itza. This idea was based on the interesting stories that my old college friend, Jim Westphal, had told me about his days working on an oil company seismograph crew in that region. Jim had suggested that I make a trip someday from Mexico City around the Gulf of Mexico to Yucatan by the local mail flight. Merle agreed that a trip to Yucatan sounded interesting. I bought aeronautical charts of the flight path around the Gulf of Mexico and started working out the details for the side trip.

On September 30, Merle left St. Louis for Dallas, Texas, on the Missouri-Pacific Railroad's *Texas Eagle*. The next day I drove down from Ardmore and met Merle at the Dallas airport. We then boarded an American Airlines flight, and 1,000 miles later we arrived in Mexico City. We took a taxi into town and checked into our modest hotel room. I complained to the management about the prominent ring around the bathtub, and someone came to our room and scrubbed vigorously, but to no avail!

We took a city bus tour on our first day in Mexico City. Our guide told us that Mexico City was founded in 1325 AD by the Aztec Indians on an island in the middle of a lake. According to legend the Aztecs chose this spot because they saw an eagle eating a snake atop a cactus here and considered this as a sign from their gods. In the 1500s the Spaniards filled in the lake and greatly expanded the city. Mexico City is located in a valley on a high volcanic plateau. With an altitude of 7,350 feet above sea level, the climate is like spring all year around. Our tour visited all the important tourist sites, including: Alameda Park with its fountains; the Monument of Benito Juarez, one of the greatest heroes in Mexican history; and the Government Communications Building, covered 10 stories high with colorful murals of Mexico's history and workers. The downtown area impressed us with its well-dressed people, clean streets, and orderly traffic, but the rest of the city shocked us with its poverty and terrible traffic congestion.

Our bus took us to a high point where we could view the distant snow-capped volcano, Popocatapetl, which rises to 17,887 feet, and its sister volcano, Iztaccihuatl. We then visited the cathedral known as the Basilica of the Virgin of Guadelupe, Mexico's most important religious shrine. We were startled to see poor peasants crawling for miles on bloody knees as a form of penitence so they could enter the shrine and donate their pesos. The shrine, built in 1695 on the site of a holy vision, was noticeably tilted to one side due to subsidence of the ground.

As our bus drove around the city, we were amazed to see that the bases of the many statues were now some 8 to 10 feet off the ground and supported only by long pillars under the bases. The area was originally a lake until it was drained by the Spaniards and consists of soft lake sediments sitting on thousands of feet of volcanic ash saturated with fresh water. As water has been pumped out of the underlying ash, the city has been sinking dramatically and alarmingly. Huge buildings, such as the Opera House, had sunk so far into the ground that former first floors were now below ground.

Then we were off to the famous floating gardens at Xochimilco about 10 miles south of town. Here we traveled by boat down canals past small islands with tall, slender juniper trees. Boats with flower-covered roofs and decks floated by selling souvenirs and food, and some boats had *mariache* bands playing Mexican folk songs. It was all very colorful and festive. Next our bus took us to a street outside of Mexico City lined with silversmith shops selling silver goods of every kind. I bought a solid silver belt buckle with my initials engraved on it and ordered a set of silver wine goblets to be custom-made to take back to Mother in St. Louis. Merle bought a silver bracelet and a purse for his mother. That evening after the tour Merle and I dined at Sanborn's Restaurant in Del Prado Square, the most famous restaurant in Mexico. The waitresses were dressed in colorful folk costumes. We asked for tacos and enchiladas and were informed that only peasants along the Texas border ate food of that kind! The Spanish-style cuisine was very tasty, however.

After dinner we went to the Fronton Mexico to watch the *jai-alai* games. For this game, two or four players in an enclosed, basketball-sized court wore long, scoop-shaped wicker baskets, called *cestas*, strapped onto their throwing arms. As a hard rubber ball ricocheted off a wall, they caught it in their baskets and hurled it against another wall with terrific speed and force; this form of handball has the fastest moving ball of any sport. As each point was made, there was a frantic round of betting with the bookmakers throwing betting slips inserted into slits in tennis balls into the highly energetic crowd.

The next day we hired a taxi driver, who doubled as our English-speaking tour guide, to take us to the bullfights at the Plaza de Toros arena. The driver entered with us free of charge and sat behind us so he could explain all the bloody action. We watched a number of young, black bulls get stuck with darts, speared, and finally killed by a matador's sword driven from the back of the neck into the heart. But the bulls got in their jabs, too. One bull worked its horns under the padding of a horse and threw it over on its side on top of the picador. A matador named Navarrito was tossed into the air by a bull and then gored on the ground. There definitely was "blood on the sand!" It was not a spectacle for the squeamish and weak at heart.

About this time we took a chance on eating some enchiladas from a street vendor who had a cart with a glass case filled some six inches deep with greasy meat. The next day, Merle came down with a bad case of food poisoning, known to tourists as "Montezuma's revenge," that he will never forget; I escaped, but my time was coming later in the trip!

As evening approached, we decided to go to the *jai-alai* games again, but as we passed the Opera House, a lady and her daughter offered to sell us tickets for that evening's performance of *La Boheme*. We bought the tickets and found ourselves sitting in the dress circle in our sport clothes next to locals all decked out in their tuxedos and evening gowns! The lead soprano was the world famous singer, Victoria de Los Angeles, but we were handicapped by not knowing what was happening on stage because the programs were in Spanish and the singing was in Italian. All we knew was that it was a very sad story.

The following day it was time to start our adventure to the wilds of Yucatan, located about 600 miles east across the Gulf of Mexico from Mexico City and bordering Belize and Guatemala. We could have flown directly to Merida, the capital of the province but instead decided to follow my friend Westphal's advice and take the local mail run. At Mexico City airport we boarded a DC-3 airplane, operated by CMA (Compañia Mexicana de Aviacion), with its steeply inclined cabin, seats for 21 passengers and twin propellers. We then took off for Vera Cruz, 200 miles to the east on the Gulf of Mexico. After stopping at Vera Cruz, with its clearly visible offshore coral reefs, we flew almost 600 miles around the Gulf of Mexico to Merida. We made stops at Minatitlan and Villa Hermosa (both in a major oil production and refining area), the small Isle De Carmen, and Campeche before finally reaching Merida. As we flew along the coast at low altitudes towards Merida, I followed the geography on my aeronautical charts; rivers emptying plumes of chocolate-brown, muddy water into the turquoise-blue Gulf of Mexico were especially impressive.

Upon arrival at Merida we took a taxi to town and settled into a nice room in the small Hotel Colon. The rooms of the Spanish-style hotel were arranged around a small patio covered with tropical plants and flowers. The two mornings we were in the hotel, we had breakfast outside by the patio with vultures perched high on the white walls watching us. That evening Merle and I walked into town to a crowded, noisy bar to drink some wine. We sat at a table next to an open window, and a shoeshine boy talked us into sticking our feet out the window to get our shoes shined. While drinking wine and sticking our feet out the window for shoeshines, we were propositioned by a young boy who swore that his sister was still a virgin. It was all quite bizarre!

The next day we embarked on a bus tour to see the famous Mayan ruins at Chichen Itza. Our bus slowly made its way across the flat land through jungles, palm trees, and fields of *henequen* (the fibers are used to make binder twine) until we arrived at the archaeological site about 75 miles east of Merida. We entered a group of thatched huts that served as offices and gift shops, paid the entrance fee, and our guide led us out to the ruins. Our guide told us that the Mayan Indians, who had established a sophisticated empire here while Europe was in the Dark Ages, built Chichen Itza between 500 and 1200 AD. He showed us their system of writing, astronomical observatory, and their elaborate calendar carved in stone, which is still accurate. We visited a number of buildings that had been markets, steam baths, temples of warriors, the cemetery, and other important places in the ancient city. Everywhere the walls were covered with carved stone skulls and heads of Kukulcan, the serpent god.

The focal point of the ruins was El Castillo, an imposing pyramid about 100 feet high with the temple of Kukulcan on top. We climbed up the 91 steps (four sides with 91 steps each plus one step on top makes the 365 days of the year) to the top platform where we took each other's pictures seated on the *Chac Mool*, a reclined human figure carved in stone that was used for human sacrifices. Then we descended a set of stone stairs into the heart of the pyramid, where we viewed an eerie life-sized statue of a red jaguar with fierce jade eyes and long, white ivory fangs.

Next we visited the Great Ball Court, which measured about 100 feet wide by 460 feet long. Its high walls caused an amazing echo of our voices as we walked around. According to the carved hieroglyphics on the walls, teams from rival cities competed to see which one could score the most points by getting a rubber ball through a high stone

**SEATED ON A MAYAN SACRIFICIAL ALTAR AT
CHICHEN ITZA, YUCATAN, MEXICO.**
This reclined figure of a man in stone, known as the *Chac Mool*, was an
altar upon which victims had their hearts cut out by Mayan priests.

ring, in the form of two entwined serpents, without using their hands.
There apparently was quite an incentive to win the game because the
losing team had their heads cut off or their hearts cut out!

Our guide then took us to see the most famous attraction at
Chichen Itza, the Sacrificial Cenote, or "Sacrificial Well." This was a
large, natural sinkhole in the ground of circular shape about 150 feet in
diameter with 70-foot vertical limestone walls dropping down to deep,
olive-green water. Here young virgin girls were given intoxicating
drink, draped with heavy jewelry, and then thrown into the deep pool
as sacrifices to the rain god; a male guard wearing heavy armor then
jumped in to accompany her and also drowned. A trove of bones and
jewelry from these ancient rites has been dredged up from the pool. I
was particularly moved by this sacrificial well, because years before I
had read in Richard Halliburton's 1929 book, entitled *New Worlds to
Conquer,* his description of how he had jumped fully clothed into this
very well. Or, as he described it, "I took one deep breath, hung for an
instant on the brink, and then plunged, seventy feet, into the Well of
Death." Fortunately, he was able to find "a band of greenery reaching
entirely to the top" and climb out of the *cenote,* after falling back in

several times. Following a full morning at Chichen Itza, our bus took us to a small, nearby village for lunch, and then we slowly made our way back to Merida.

The next day we boarded another DC-3 for the long flight with many stops back to Mexico City. About half way into the trip I was suddenly struck with a vicious case of "Montezuma's revenge." I spent the rest of the trip in agony and running to the lavatory at the back of the plane. Back in our hotel in Mexico City that evening, my sickness finally subsided.

In the morning, we took a bus tour to the pyramids of Teotihuacan located 25 miles northeast of Mexico City. Here we walked around the excavated ruins of an ancient city of some 50,000 people that reached its pinnacle in about 100 AD. At its time it was the first great city in the New World and the center of political power and culture in this part of the world. The very large and impressive Pyramids of the Sun and the Moon dominate the site. We climbed up hundreds of rather rough steps to reach the 200-foot-high top of the Pyramid of the Sun and had grand views of the entire ancient city with mountains in the background. On the way back to Mexico City our tour visited the Convent of Alcanan, built by the Spanish in 1560 AD. The golden altar and gold-framed religious paintings above it were interesting. Back in the city, we did some last minute shopping and prepared to depart.

The next day we took a taxi to the airport and, loaded with all our purchases, we boarded an American Airlines plane bound for home. Our flight made a stop in San Antonio and then landed back in Dallas, some 10 days after we had left. Merle caught his train back to St. Louis, and I drove back to Ardmore. And, thus ended our big Mexican adventure in 1955. Merle and I still enjoy reminiscing about that trip to this day, over 45 years after the event.

8 ✳

Bayonets and Chemical Weapons

I found myself in the fall of 1956, at age 25, suddenly transformed from a geologist in Ardmore, Oklahoma, to an army recruit reporting for duty at Fort Bliss in El Paso, Texas. Eight weeks of basic training in the dry desert of west Texas lay ahead of me. Our training group was all army reservists, so I was the only 25-year-old among 17- and 18-year-old whites, blacks and Hispanics from Oklahoma, Texas, and Louisiana. I was to discover that I was in the middle of an explosive ethnic mixture. On arrival at Fort Bliss I went through the usual hair removal, uniform and rifle issuance, and bunk assignment. My fellow recruits and I then spent many weeks learning basic military procedures such as drilling, bayonet charges, polishing brass, falling out for surprise midnight inspections, cleaning rifles, polishing boots, and doing kitchen patrol (KP). During these weeks the radios in the barracks blasted with the latest songs like "Rock Around the Clock" and "The Green Door."

One day I went to the lieutenant in charge of our unit and requested permission to obtain an absentee ballot to vote in the Dwight Eisenhower

(Republican) versus Adlai Stevenson (Democrat) presidential election. The lieutenant was shocked to find that I was older than he was (he was 21). After approving my ballot request, he started looking through my records. Noting that I was a college graduate, he asked me if I had any special skills that he could use. When he found out I could type pretty well, he assigned me to do his office typing, including orders, instructions, and memos. This got me out of KP and other manual labor.

Life in our barracks was quite an eye-opener to someone like me, who had led such a sheltered life. There was continuous fighting between the working-class whites, blacks and Hispanics. Our barracks's billiard room looked like a battle zone with holes and slashes in the walls from fighting with pool queues. The year was 1956, years before any civil rights laws came into effect, so we had our own mini-race riots right in the barracks.

As our training progressed, we were issued weekend passes. Some of us went into nearby El Paso, but most of the recruits rushed straight across the U.S. border to Juarez, Mexico (called "Wa-zoo" on the base). The bars, nightclubs, and girls there were the greatest things these 17- and 18-year-olds had ever seen. Some of our guys even went AWOL (absent without leave) from our base to visit Juarez and were hauled back by the military police. I visited Juarez a couple of times for sightseeing, shopping, and drinks in a couple of seedy looking bars, but the scene really didn't appeal to me. At one bar I asked the Mexican bartender why the beer was so expensive and was surprised when he said the cost included the services of the bargirls!

The culmination of our eight weeks of training was a one-week bivouac (camp-out) at the Kastner firing range located in the desert about 20 miles north of our barracks. We were trucked out to the campsite, set up individual "pup" tents on the ground, and ate at a field kitchen. As it was getting into winter, the days were sunny and nice, but the nights were bitterly cold. Every night I slept in my little pup tent on a ground sheet inside a sleeping bag with three wool liners but still shivered all night with the cold. These were some of the coldest nights of my entire life. We spent most of the days of our bivouac out on the firing range with our M-1 rifles, firing at different targets at ranges from 50 to 500 yards. No one, including the officers in charge took the scores seriously as it had already been decided that everyone would qualify for his

"Sharpshooter Medal." As a result, no scores were recorded and everyone fired at whatever targets were up and particularly enjoyed shooting at the red marker spots every time one was held up in front of someone else's target. The shooting was lots of fun until we heard that a high-ranking officer was coming to inspect the range and watch the shooting to see how a reserve unit performed. Our lieutenant ordered me to fill up the entire scoreboard for all names and all ranges with hundreds of fictitious scores for the visiting officer to inspect. That's when I found out how hard it is to think up random numbers; no matter how hard I tried my numbers kept falling into patterns.

Our activities at Kastner rifle range also included running through an obstacle course with live machine-gun ammunition being fired overhead. I smuggled a cheap camera that I had bought in El Paso on to the range and captured many bizarre scenes from our bivouac. One morning we were all sitting by our tents shaving when a very naive recruit sitting next to me kept complaining that no shaving soap was coming out of his aerosol can. The next thing I knew, he had stabbed the can with his bayonet and a cloud of shaving cream rose up in the air and then settled all over us. I still have the photo to prove it.

BASIC TRAINING AT FORT BLISS NEAR EL PASO, TEXAS.
That's me on the left with a fellow recruit (the one that stabbed his can of shaving cream). Learning how to become "trained killers!"

The last day of our bivouac we were ordered to pack up all our gear and get ready to return to barracks. We were then shocked to learn that, instead of returning in trucks, we had to march the 20 miles back to camp on foot with full pack. It was a long, hard hike and we stumbled into our barracks in the middle of the night totally exhausted. And then, after eight weeks of marching, training, cleaning and polishing, firing weapons, race riots, and other sordid activities, we marched through our final review and were given the rank of Private E-2. We were then each issued separate orders as to where we would spend the final four months of our reserve active duty training. Most men went into the infantry, but because of my "scientific and professional personnel (SPP)" rating, I was assigned to the headquarters of the Army Chemical Corp. at Edgewood Arsenal, Maryland, on the western shore of Chesapeake Bay. This all sounded like a very intriguing and interesting assignment for me.

I flew back to Tulsa, got my MG sports car out of storage, and headed east in the middle of winter. I stopped off in St. Louis to visit Mom and Dad and then started a long, cold drive east to Maryland. My little MG had what I called a "psychological heater"; it made a lot of fan noise that made me think I was getting warm but actually it put out very little heat. With little heat and cold air blowing in around the loosely fitting, removable window panels, it was freezing in the car. I drove with a blanket wrapped around my legs and wore a coat, gloves, hat, and scarf as I drove across the snow-covered terrain of Illinois, Indiana, and Ohio. I reached the Appalachian Mountains in West Virginia and climbed slowly up the mountains in low gear through snow and fog. At one point my windshield got so iced up that I took the left window panel out and leaned out of the window to watch the road. Suddenly a car coming downhill shot past me and showered me and the interior of the car with wet, slushy snow and black cinders. I had to wipe slush off my face and clean it out of my left ear! Coming down out of the mountains, I found my way to Edgewood Arsenal on Chesapeake Bay, and reported for duty. Edgewood Arsenal, also known as The Army Chemical Center, was originally built during World War I to make poisonous mustard and phosgene gas, and during World War II made 63 different chemical weapons, including the incendiary bombs that General Dolittle dropped on Tokyo in 1942.

When I arrived at Edgewood, I found that most of the army personnel on the base were college graduates with science backgrounds who were assisting civilian scientists do research on chemical, biological, and radiological warfare; there was also a small cadre of regular army types who were assigned to look after us. I was pleased to find that most of the other men had been drafted for two years while my SPP rating meant I had only four more months to go.

At Edgewood they asked about my special skills and found out that I could operate a Friden electric calculator. This large desktop calculator did complex calculations by mechanically jumping a carriage back and forth across an adding machine. I later learned how to do the "Friden March"; when I entered a certain calculation, the machine would whirl and bang in the tempo of a march.

I was assigned a one-man research project, which centered on a large concrete gun emplacement, or bunker, located in a field surrounded by woods. The bunker windows and interior were fitted with a number of light beams and electric eyes to measure light intensity, and these were wired to a circular, multiple-pen, paper-chart recorder. Every day I would measure the wind direction and velocity and then smoke up the interior of the bunker using a beekeeper's tobacco smoker while wearing a gas mask. This smoke simulated poison gas from a rocket fired into the bunker. I would then switch on the recorder and wait outside until all the smoke had blown out of the bunker. Following each field operation, I took my data back to my office and did calculations on my Friden calculator. I then plotted the results on a chart. While I was out smoking up my bunker, my friends at the base were in laboratories slopping around with deadly nerve gas (one drop on the skin was lethal), testing flame-throwers, or doing other interesting weapons research.

The military aspects of our life at Edgewood were unique and quite humorous. All week we were scientists, but every Saturday morning we gathered for army assembly, inspection, and instruction. We college boys loved to joke around with the few regular army types on the base and they seemed to discreetly enjoy the fun. For example, when our sergeant would order us to "march," everyone would stomp their feet and "moo" like cattle, and then we would finally march off. Working in laboratories all week also gave us time for mischief. One time, when official photographs were posted in our barracks showing how foot and wall lockers should be neatly made up, one of the guys working in a photo lab replaced them with photographs of messy lockers complete with photos of nude women.

Another game we played with the regular army types was in the form of an imaginary anti-military fraternity called the "Phi Tau Alpha," where the Greek letters stood for "Phooey (actually a much stronger "F" word) to the Army." Every time the army staff would post notices or instructions, someone would write "ØTA" across them. This game finally went too far, and the army officer's called in the Criminal Investigative Division (CID) to find out who was defacing official government notices. We heard they were going to dust everything for fingerprints, so one of our guys defaced a notice and then used a monkey in one of the laboratories to put monkey paw prints all over it for the CID to find.

For recreation on weekends I would drive my MG to New York City where the United Services Organization (USO) fixed me up with free theater tickets (as long as I was not in uniform), cheap hotels, and inexpensive food. I particularly enjoyed bars with Dixieland jazz bands in those days. The trips were great fun and I got to know the Times Square area quite well; it was relatively safe walking around New York in 1957.

My good friend and fellow geologist, Mal Boyce, and I also enjoyed going into Baltimore (pronounced there as "Ball-a-mer"), Maryland, where we would sometimes visit the Gaiety Burlesque Theatre. The Gaiety was a seedy old place, and the shows consisted of the usual dreary strippers alternating with old vaudeville-style comedians with rude and crude jokes. A vendor, who was obviously a punch-drunk ex-boxer, would come down the aisles selling ice cream bars and wiping his nose with his hand holding the ice cream, which Mal and I thought was very humorous. We usually went to the Saturday matinees, when the few customers were weirder and more humorous to us than the comedians. In Baltimore, we also frequented good seafood restaurants and a few old, dark-paneled bars.

On weekends, I also enjoyed driving my MG around exploring the 200-mile-long Chesapeake Bay of Maryland. The small, picturesque fishing ports were fun to visit, and the seafood was great too, especially the soft-shell crab sandwiches. One weekend I drove to some cliffs on the bay near the U.S. Naval Academy at Annapolis, Maryland. These cliffs are famous for the large fossilized shark's teeth that weather out of the cliffs and are found on the beach below. I searched around for an hour or so, but did not have any luck. As I toured around Chesapeake Bay, I was fascinated by the fleet of working, 40-foot-long sailing boats called "skipjacks." These low-cut, fishing sloops had strongly raked (tilted back) masts, two massive triangular sails, and a prominent spar projecting forward of the sleek bow or cutwater. I never tired of watching these sloops sailing along under full sail as they dredged up blue-point oysters.

But, all good things must come to an end. In April 1957 my six months of active army reserve duty were completed. I packed up my MG and headed back to Tulsa, Oklahoma, to return to work as a geologist with the Ohio Oil Company. Looking back, I had enjoyed four months in Maryland with my army buddies and had experienced many memorable visits to New York City, Atlantic City, Baltimore and Chesapeake Bay. My days at Edgewood Arsenal were probably the most relaxed in my life as everything was routine with no real responsibilities to worry about.

9 ✴

Assignment Libya, North Africa

In April 1957, my active army reserve duty had been completed and I headed back to Tulsa, Oklahoma. While in Tulsa I visited the Ohio Oil Company Division Office and was told I had been selected to go to Libya to do surface geological surveys in the Sahara Desert. This was what I had been waiting for since high school! The Tulsa senior management told me that the home office in Findlay, Ohio was impressed with the answers on my questionnaire requesting assignment to Libya, especially my signature in Arabic. They flew me to Findlay to be interviewed for Libya by the Chief Geologist, Phil M. Konkel. After a friendly and positive interview, Konkel asked me if I could read the nameplate in Arabic on his desk. I read it as "R.N. Konkel," which was confusing as these were not his initials, but then he smiled and said, "That's right, that nameplate was actually made for my brother but they made a mistake in the last name so I kept it." Needless to say, Konkel was impressed and I got the job.

I made a brief, final trip down to Ardmore, Oklahoma, to say goodbye to the office staff and my other friends, and then moved back in with my Aunt Peg and Uncle Marian Halsey in Tulsa to get ready to go overseas. I sold my MG sports car for $775 (parting with my MG was rather sad), traded all my amateur radio equipment for a good short wave radio receiver, and bought warm-weather clothes for Libya. I packed my duffel bag with the full kit of army clothes that I had retained because I was to be in the inactive reserves for the next six and a half years. Little did I know at that time that I would be living in my army-issued khaki uniforms, boots and field-jacket for the next three years in the Libyan Desert.

While I had been working for Ohio Oil in Ardmore, Oklahoma, I had eagerly watched developments in Libya with plans to go there as soon as the company would let me. The Government of Libya was working hard to pass its first petroleum law in order to grant oil exploration concessions to foreign oil companies. The country was invaded by Italian forces in 1915 and became an Italian colony, but Italy lost it as a result of having joined Germany during World War II. In 1951, the United Nations proclaimed Libya a monarchy under King Idris. As a sandy desert nation without significant agriculture or industry, Libya was very poor and desperately needed oil income to get the new country going.

Libya passed a law in 1951 allowing reconnaissance geological surveys to explore for signs of oil, but no wells or geophysical seismic surveys were permitted. Under this law, Ohio Oil had joined with Continental Oil and Amerada Oil to form an exploration company called Conorada (a combination of the three company names) Petroleum, which began reconnaissance work in Libya in 1954. Three Conorada field parties made geological surveys all over Libya during 1954 and 1955. Their biggest problem was the millions of World War II landmines scattered all over the desert and still quite lethal due to preservation by the dry desert climate. The crews had a strict rule to always travel in tire tracks in the sand made by previous vehicles. The Conorada geological parties found that Libya had all the requirements for major oil discoveries, including large anticlinal structures and a thick section of sedimentary rocks with both porous reservoir layers and oil source rocks.

Libya passed its new Libyan Petroleum Law in 1955, and later that year the Conorada Group was awarded Concessions 25 through 33, totaling 34,876,011 acres, or almost 55,000 square miles. In 1956, Ohio Oil was made operator of the concessions using the name, "The Oasis Oil Company of Libya." In 1957, Oasis Oil was also awarded Concessions 59, 60 and 71, bringing the group up to its maximum acreage position of 62,032,493 acres, or almost 97,000 square miles. The thirteen Oasis concessions, located all across the northern part of Libya, were well spaced in an east-west direction from the Egyptian border all the way to the Tunisian-Algerian border.

Oasis started surface geological, seismograph and aerial photograph surveys in 1956. And, now I was going to join this large and important operation.

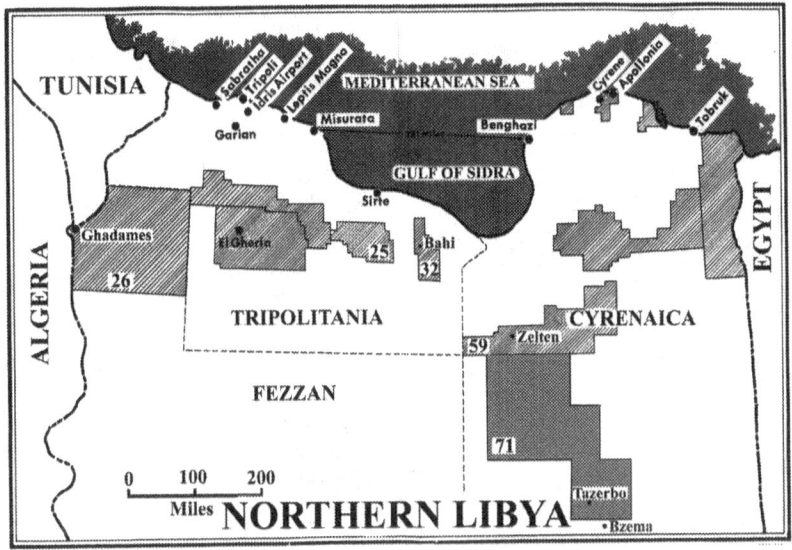

INDEX MAP OF NORTHERN LIBYA SHOWING CITIES, TOWNS, OASES AND PETROLEUM CONCESSION AREAS OF THE OASIS OIL COMPANY OF LIBYA (ABOUT 1958).

On May 9, 1957, I flew to New York City and stopped over to get my yellow-fever shot at the U.S. Health Department facility in The Battery. Then, I boarded a British Overseas Airline Corporation (BOAC) DC-6 with four propeller engines and flew at 300 miles per hour to Boston and then across the Atlantic Ocean. During the twelve-hour flight across the ocean, my first Trans-Atlantic crossing, I sat in First Class next to an elderly British lady from Wales. For dinner, the steward laid out eight pieces of silverware for me, and I said to my flight companion, "You must think we Americans are very barbaric to eat a meal with only three pieces of silverware." She replied, "No, I think it is very clever of you!" The next morning we landed in Prestwick, Scotland, just in time for breakfast to be served in the BOAC hotel dining room by waiters in white gloves and claw-hammer coats. After breakfast, I cleared through British customs where it was discovered my suitcase was missing. Then I flew on to London, where my suitcase was still missing. Several days later my suitcase was found and BOAC said it would be forwarded to Tripoli.

On May 13, I boarded a BOAC aircraft and flew at 18,000 feet at 260 miles per hour south to Tripoli, Libya, passing over the snow-covered Alps and the blue Mediterranean Sea on the way. It was night by the time my plane flew over the city of Tripoli on the North African coast of Libya. All I could see was a scattering of dim yellow street and house lights. Then, we left the lights behind and headed out over the dark desert. Some twenty miles later, we landed at 10:30 p.m. at the airport. All was dark except for a group of dimly lit buildings some distance away. Several armed soldiers escorted us to a large Quonset hut full of bullet holes from World War II where we went through customs and immigration procedures.

I was surprised and concerned that no company representative was at the airport to meet me but then realized that we had landed an hour or more earlier than scheduled. I boarded a bus that took me to the Hotel Uaddan in Tripoli, where most foreigner visitors stayed. During the bus trip at midnight, a full eclipse of the moon occurred and I wondered at the time what kind of omen this foretold. Two hours later, several Libyan employees of The Oasis Oil Company found me sitting in the hotel lobby and took me to the company's bachelor quarters for geologists, where I moved in with Warren Heisterkamp. Warren was known to us as "Doc," because he had been asked by a Bedouin chief in the desert to circumcise his son and, finding no way to get out of it, he performed the operation having had no medical experience whatsoever.

The next day, May 14, the weather was beautiful and one of the Oasis bachelors, a Canadian named Jack Redmond, with whom I had gone to college, took me on a tour of Tripoli. We visited the crowded and noisy markets of the Old City, the old Turkish fort, the business district, sidewalk cafes, and other sites. I was very impressed with this interesting, historic, and rather clean city.

THE HARBOR FRONT OF TRIPOLI, LIBYA IN 1957.
This beautiful palm-lined harbor front is now reportedly
filled in and lined with warehouses for the oil industry.

The next few days were spent meeting the Oasis managers and office staff, looking at geologic maps, visiting the warehouse and garage, and attending a cocktail party in my honor. At my cocktail party, one of the geologists told me that he was in the desert when the recent eclipse of the moon occurred. He said his Libyan workers got all excited and then a guard took a rifle shot at the moon as it started to re-emerge into view. It seems that the Libyans believed that a lunar eclipse was a whale swallowing the moon and if a lot of noise were not made to scare it away everyone on earth would die.

The weather continued to be perfect. I was introduced to the Underwater Explorers Club located west of Tripoli. I spent a lot of time snorkeling around the reefs looking at the fish and pieces of Roman pottery scattered all around in the clear, warm water; this was sure better than dunking in the murky water of Lake Murray back in Oklahoma. I also drove around the outskirts of town and saw the American Wheelus

Air Force Base; British army barracks; World War II gun emplacements and many battle-scarred buildings, some with "VV IL DUCE" (long live the Duce, Mussolini), written on them.

I ate all three meals a day with the other Oasis bachelors at various Italian restaurants. The waiters spoke pretty good English but got mixed up at times. One day, I asked an Italian waiter if they had ice cream. He said, "Yes." I asked if it was pasteurized. He said, "Yes, we have apple pie." I said I meant to ask if they had Olympia ice cream, the only one that was pasteurized. He said, "No, only vanilla!" I never did get any ice cream.

By May 27, I was getting ready to make my first field trip into the Sahara Desert to Oasis Oil Company's Petroleum Concession Number 26 on the border with Tunisia and Algeria. However, a severe sandstorm, or *ghibli*, had blown in from the desert and the field temperatures had gone up to 115 degrees with blowing sand. The weather in Tripoli turned hot, dry and dusty. News arrived that a typhus epidemic had broken out in Benghazi in eastern Libya where Oasis also had an office. Several of our geologists needed typhus booster shots so it was arranged to fly some vaccine to them on a U.S. Air Force radar plane that regularly flew from Wheelus Air Force Base over to Benghazi. The ghibli continued to blow in the desert for ten days with 50 mile-per-hour winds, blowing sand and dust, and temperatures reaching 120 degrees. My first trip to the desert was postponed until the weather calmed down. Finally the weather cleared and I flew with Oasis geologist, John Dunlap, in the company's twin propeller C-47 cargo plane (the military version of the DC-3) to the large oasis of Ghadames, located near the junction of the Libyan, Algerian and Tunisian borders. As we circled for landing, I could see the old French Foreign Legion fort, large groves of date palms and a city of small buildings painted white with each corner turned up into a point. I had finally made it into the Sahara Desert!

While parked on the dirt airstrip at Ghadames, we unloaded our equipment and food from the plane. Then, the mayor of the oasis, or *mudir*, suddenly arrived and asked our pilots to take one of his relatives with a broken leg back to Tripoli for medical attention. Our pilots were very reluctant to get involved but finally consented. An ambulance arrived and a woman covered from head to toe with a native robe, or *barracan*, was put on board our plane. As the plane prepared to return to Tripoli, John Dunlap and I went into town to a little store and had a slightly cool bottle of orange drink. A man sat in a corner smoking a water pipe and a group of Tuareg tribesmen stood around outside. We walked around the ragged and dirty Tuareg camp just north of town. The Tuaregs, known as the "Men of the Blue Veil" because of their indigo colored robes and veiled faces, were once known as fierce fighters. They had once become wealthy by demanding protection money from all the caravans coming up from Central Africa, but, this life had been over for many decades and the once proud Tuaregs now lived in poverty and squalor and ignored by the government.

Dunlap and I picked up the Land Rover he had left parked at the airstrip and drove east to his camp some 20 miles away. When we arrived, it was 110 degrees in the tents, but at sundown the temperature dropped sharply due to the low humidity and it seemed cool. A team of two other geologists, Red Swanson and Bill Noonan, was there to greet us. Dunlap showed me my one-man tent, which had a double canvas roof as insulation from the sun. Inside were a canvas cot, inflatable air mattress, sleeping bag with pillow, and a footlocker. I stored my personal effects in the footlocker and found that the personal servant, or "batman," had placed a pitcher of water and basin outside my tent on an orange-crate. I washed up and went into the small trailer, which served as both dining room and office and had a small and very hot kitchen at one end.

We were served dinner by the Libyan cook and called Tripoli on our short-wave radio to report we had safely arrived in camp. After dinner, I hooked up my short-wave radio receiver in my tent and listened to the news from the British Broadcasting Corporation (BBC) in London before retiring on top of my sleeping bag. Our electric generator was shut down at 10:00 p.m. for the night; the camp became dark and quiet. At four o'clock in the morning in the middle of a warm, quiet night, the wind suddenly shifted strongly into the north blowing sand and dust all over me as I slept on top of my sleeping bag. The north wind cooled things down.

Dunlap, Swanson, Noonan and I spent all the next day getting the camp in shape. Two large tents had been torn in half by the earlier sandstorm and had to be repaired. The vehicles were serviced and checked out. At nine o'clock in the evening a truck from a Robert H. Ray geophysical crew came into our camp. The Libyan drivers explained

**ME AND MY TENT IN THE SAHARA DESERT OF LIBYA NEAR
GHADAMES OASIS.**
The tent had a sunroof and mosquito netting. Washing-up
facilities included a "jerry-can" of water and a basin. Also seen
is a canvas bag to cool drinking water by evaporation.

that a water truck had left their camp several days ago to fill up at a nearby oasis but had disappeared. They had just found the truck not far from
our camp but there was no sign of the drivers. We told them we had not
seen them.

As we prepared to bed down for the night, we noticed the kerosene-
fueled refrigerator full of meat was starting to smell bad. Obviously, the
small kerosene lamp inside the fridge was not vaporizing the refrigerant to
cool the refrigerator down. We decided that these British-made kerosene
refrigerators might work fine in East Africa, but not in the Sahara
Desert. The next afternoon, our water truck returned from Ghadames
oasis and we treated some of the water with Halizone (chlorine) tablets
for drinking.

The next day started out quite cool, as the wind had shifted into the
north. Dunlap and I headed into the field in our Land Rover to map
some geology, but we found it was too hazy to use our survey instruments.
We drove around the desert doing reconnaissance as the temperature
slowly climbed. After a box lunch, we went back to camp. Swanson and

Noonan had left for the oasis of Dirj for gasoline. The weather turned very hot again and we could only map geology early in the morning and late in the afternoon. The heat waves made instrument readings impossible during the middle of the day and the breeze often died down at that time of day too. While lying on our cots in mid-afternoon, the temperature reached about 115 degrees in our tents. At night the low was about 80 degrees at four o'clock in the morning.

We heard on the company radio that a gasoline truck had gotten lost traveling from Tripoli to a Robert H. Ray seismic camp south of us, but it had finally wandered into their camp five days later. Then, they found the tank had sprung a leak and all the gasoline had leaked out!

On June 17, Dunlap and I set up a survey station on a butte. Looking north across a salt flat, or *sebcha*, we could see in the distance a small Foreign Legion fort at the foot of a sea of very large sand dunes. Dunlap said the fort was called Mzazzem and was very close to the Libyan-Tunisian border. Dunlap recalled that several months earlier our fellow geologist, "Doc" Heisterkamp, was flying reconnaissance over this area and spotted the fort of Mzazzem. He saw trucks parked around the fort and out of curiosity he had his pilot circle and land. As they walked up to the fort, several men came running out and yelled that they were filming a movie and they had just ruined their shots by flying over the fort. Heisterkamp went inside the fort and found John Wayne and Sophia Loren acting out a scene with a string of donkeys. John Wayne was swearing and kicking the donkeys in frustration as they filmed scenes in the heat and flies for a movie called, "Legend of the Lost."

That night, as I tried to sleep in the heat, a sandstorm blew up from the south and the temperature went up over 100 degrees. Quartz sand grains blowing through my tent at high speeds made sparks as they hit the metal zipper of my tent's mosquito netting. The sandstorm continued to blow for several days making geological surveying almost impossible and driving temperatures up. The next day, the Oasis single-engine airplane, a DeHaviland Beaver, managed to briefly land on our landing strip, despite the high winds and blowing sand, with a shipment of fresh meat and mail. We tried to sleep in the afternoon heat and go to the field for a few hours in the evening. One afternoon, our driver, Ziglam, got up after a nap, and still half asleep, drove through a guy-wire pulling down our radio antenna mast.

In the evening after work, I would take a shower in our little shower tent to wash off the sand and to cool down. This involved working the handle on a pump stuck into a drum of warm water. After wetting down, I would grab a handful of powdered soap detergent (it was rather rough on the skin but was the only thing that would lather in the slightly salty water), work up a little lather, and then continue pumping until finished. Finally, the weather turned cooler as the wind again shifted into the north. There were even a few thunderstorms and day temperatures only got to 106 degrees. We took advantage of the cooler weather to map a number of geological survey points to the east of Ghadames oasis.

On June 21, we went for supplies in the small oasis of Dirj (also spelled Dirdj and Dirg, as there is no way to translate from the Arabic into English except phonetically. In one of Lawrence of Arabia's books, the editor records that he complained to Lawrence that he spelled his camel's name as Leila in one place, as Layla in another place, and Lila in another. Lawrence's reply was simply, "Ah, yes, she was an excellent beast!"). Dirj consisted of a bleak group of mud buildings with a few date-palm trees. As we drove in, dirty little Tuareg children laughed and shouted the Tuareg greeting of "Le-buss" to us. Veiled men and unveiled women with silver disks on their foreheads came out to greet us. After buying water, gasoline and a few other supplies, we were invited up some steps into a mud-brick room to have tea and orange soda pop. The shopkeeper and a couple of his friends had great fun joking around with us as we drank their tea. As we drove back to camp, we had two flat tires and got stuck in the sand several times requiring steel sand-tracks and winches to get our vehicle out.

Our geological surveys continued for about two weeks and then it was time for us to go back to Tripoli for "R and R" (rest and relaxation). We drove west through several large locust swarms to Ghadames oasis. We walked around and saw water wells, aqueducts, date palms, and the Tuareg village while we were waiting for the Oasis C-47 to arrive from Tripoli. Our plane arrived and on board were executives from the Ohio Oil Company headquarters in Findlay, Ohio, and our Chief Geologist, John Goffe, from Tripoli. Dunlap and I joined their party to take a tour of Ghadames.

I was to find out in the weeks ahead that Ghadames was the most fascinating oasis in Libya. It is located 250 miles south of the Mediterranean coast and when the Romans were here it was their most southern outpost; their water system was still distributing water all over the oasis from a deep spring surrounded by palm trees. Most of the 3,000 inhabitants live underground to escape the desert heat, but also have one-story houses above ground. Following our tour of Ghadames, we flew back to Tripoli on the company plane.

THE "SPRING OF THE MARE" IN GHADAMES OASIS.
The water distribution system in this desert outpost was built by the Romans some 2,000 years ago and was still supplying the oasis.

I spent the rest of June and the month of July in Tripoli because it was too hot to work in the desert. The climate in Tripoli was pleasant even without air-conditioning. Our office was on "summer hours," which meant we only worked two afternoons a week. This left us five afternoons a week for personal activities. I joined the Underwater Explorers Club, located just west of Tripoli; this club was run by an Englishman named Simon Codringham, a leftover from World War II, and had a rocky beach, deep and clear water, and served good meals. My bachelor friends and I also did lots of snorkeling and spear fishing west of Tripoli and saw all kinds of fish, Roman pottery and debris from World War II.

We also explored the magnificent Roman ruins of Sabratha and Leptis Magna located near Tripoli. These ancient cities dated back to 1000 BC under the Phoenicians but reached their zenith during the reign of Roman Emperor Septimius Severus (193-211 AD), who was born in Leptis Magna. These ruins were more spectacular than many in Italy because the Romans abandoned them to the sand rather than tearing them down to use the stones for new buildings, as had been done in Italy.

THE THEATER AT LEPTIS MAGNA.
The Roman city of Leptis Magna flourished under the
Emperor Septimius Severus (ruled 193-211 AD).

After work on many afternoons, fellow geologists and I sat at a sidewalk café and drank beer and ate pistachio nuts while watching all the local traffic going by in horse-drawn carriages and on donkeys. In the evenings, my bachelor friends and I dined very well in Italian restaurants and went to the local movie theaters. Some movie theaters were in English and some in Italian; I remember seeing "The Bridge on the River Kwai" in Italian (the Japanese General would say, "Buon Giorno" when greeting his staff). I bought a brand new Volkswagen "Beetle" automobile, which was nice for sightseeing trips around the Tripoli region. Life in Tripoli was indeed pleasant and interesting.

On August 3, 1957, John Dunlap and I flew back to Ghadames on the company C-47. We found the Dodge Power Wagon that our other crew had left for us at the airport. The truck had a flat tire and no jack so we borrowed a jack from the ever-present policemen. We then drove east to our campsite and got everything ready to go back to our geological surveys. The next morning we were pleased to find the weather pleasant and not too hot.

For several days, Dunlap and I surveyed in a number of geological observation points. These points, which we marked with a pile of rocks, called a *cairn*, were on top of buttes several miles away from each other. Looking through the telescope of our survey instrument we could just barely see our Libyan assistant holding up a 15-foot survey rod at the other survey point miles away. The distances were so great that we had to make a correction to our readings for the curvature of the earth. The heat waves limited our work to early in the morning and late in the afternoon. Once a point was surveyed in, we would then describe the geology at that station for our maps.

SURVEYING GEOLOGY IN THE DESERT.
Here I am out on reconnaissance using my binoculars and a
Landrover automobile. Most of the Sahara Desert is rocky
terrain like this rather than seas of large sand dunes.

Dunlap and I made another trip to the oasis of Dirj where we bought drums of gasoline, kerosene, and diesel fuel. The owner of the fuel shop, Abu Dhat, again invited us up to his rooms for tea and the local unleavened bread. The next day, Dunlap and I left camp in a

Dodge Power Wagon with another Power Wagon following us (we never went anywhere with only one vehicle). We made a reconnaissance trip south of Dirj into the Wadi Tanaret area (a *wadi* is a low place or dry riverbed in the desert) to find a new place for our camp and mark out an airstrip in the sand. We found that we had to cross a sand dune-filled wadi between our camp and the area we were scouting. One of our trucks got stuck and had to be winched out with the other truck. We made several such trips and finally located a new campsite where we could extend our surveys to the east. The afternoons were getting very hot with temperatures over 100 degrees.

We were preparing to move camp when we heard that the Oasis C-47 was coming the next day with supplies. So, Dunlap decided that this was the day to move camp. We started tearing down our tents and packing up. We went over to a nearby Robert H. Ray Seismic camp (they were working for Oasis) and borrowed a Ford F-800 truck to help with our move. Then, without asking permission, our driver, Ziglam, jumped into the Ford truck, ground the gears, and broke the transmission. The Ray Camp brought over another truck and finally at 4:30 p.m. we were ready to move out. Our convoy consisted of: three Dodge Power Wagons, a Ford Tandem truck, a Ford F-800 truck, our office/kitchen trailer, a water trailer, a portable generator, and a Land Rover. Our convoy got underway and almost immediately the rough road caused the filler cap on the portable generator to break off and diesel fuel splashed all over it. Then, the convoy hit the wadi full of long sand dunes. Although this wadi was only about half a mile wide, it took us nine solid hours of winching, towing, and using metal sand tracks to get across it. During this ordeal one of our Libyan drivers kept the starter going on the stalled Ford F-800 truck until flames burst out of the engine. Another driver tore up the winch on his truck.

We finally arrived at our new campsite at four o'clock the next morning. We set up our cots and slept under the stars for about two hours. After a quick can of consommé for breakfast, John Dunlap and I fired up the short-wave radio to call Tripoli but found that the transmitter had been damaged during the move. The C-47 was due to arrive from Tripoli with supplies at about 10 a.m., so John and I headed out for the airstrip. As we approached we saw our plane take off and fly away! We asked a policeman standing nearby what had happened. He said the plane had arrived early and not found us and so it was going to Ghadames and then come back to Dirj. We waited in the heat for three hours but the plane did not reappear so we returned to our new campsite and started setting it up. That night we saw a bright and very impressive comet in the northern sky; we wondered if this was a sign of better, or worse, things to come for our camp move.

At 7 o'clock the next morning we heard John Goffe on the radio telling us that our supplies had been left in Ghadames, but we could not reply as our transmitter was still out. Dunlap and I drove our Land Rover over to a nearby Robert H. Ray seismic camp and used their radio to request that a radio repairman and fresh meat be sent out to us on the DeHaviland Beaver aircraft. Then, we started the 45-mile drive to Ghadames. On arrival at Ghadames we looked up a Frenchman, a Foreign Legion veteran, who looked after the airstrip. He showed us our supplies and then Dunlap decided to give all the meat to him, as it would surely spoil before we could get it back to our camp. He was very pleased with this bounty and invited us back to his rooms in the old Foreign Legion fort for a cool beer with his colleague, a French meteorologist. It was really hard to comprehend how these two rather cheerful Frenchmen could live out in this remote Saharan oasis.

On return to our campsite we repaired our generators and prepared a rough landing strip for the Beaver to use when it arrived the next day. At 11 o'clock that night after the generators had been shut down, John Dunlap stepped out of his tent with a flashlight and thought he saw something move. He discovered it was a deadly horn viper only six feet away from him. He killed the snake with a rock and it measured over two feet long. The next morning the Beaver arrived from Tripoli with our Italian radio repairman, meat and mail. Italians do not do very well in the desert, and soon our radio repairman got sick while working on our radio and urgently ran "over the hill" to relieve himself. After the Beaver departed we started raking rocks off of the airstrip to improve it for future landings. One of our guards threw his rake down and refused to work saying that this was not his job. Dunlap told him that if that was what he wanted then he would have to guard the airstrip all night; his bed was confiscated to make sure he did his job.

By August 18, we were ready to recommence our geological surveys. However, we could not get much done because heat waves and blowing sand restricted visibility. The flies were also very bad because of the nearby villages. Dunlap and I located and "surveyed in" several far-flung geological stations to set up our triangulation network. Travel in our Land Rovers and Dodge Power Wagon was very slow due to alternating high rocky ground and low spots filled with soft sand. We had to stop work during the afternoons due to heatwaves and the high temperatures so we lay sweating in our tents. I soon had the outline of my body in white salt crystals on the top of my sleeping bag. Suddenly at 8:30 p.m. one night we were hit by gale force winds from the north, which caused a severe sandstorm that lasted all night. I moved my cot and bedding into the office trailer for the night. The next morning the weather was delightfully cool but our whole camp was covered with sand and we had to repair tents and equipment.

The weather turned hot again but we were able to extend our geological surveys over a considerably larger area. Then we were happy to receive by truck a new trailer to use as our office, radio room and dining room. It had an air conditioning unit in it that actually worked and kept the inside temperature down to about 82 degrees, which felt very cool. Of course the little kitchen at one end where our Libyan cooks worked was not air conditioned and got extremely hot while baking was in progress. In one camp, a sudden fire in their kitchen killed their two cooks.

During our fieldwork, Dunlap had been dumping chlorine powder into our portable water tank before it was taken by our Libyan crew to a nearby oasis to be filled. We got a big shock one day when one of our Libyans happened to mention that they had been thoroughly flushing out the tank before they filled it. We had been drinking untreated water from the dirty little oasis of Dirj!

On August 23, it was time to halt our work and head back to Tripoli. Dunlap and I put together a convoy to take our equipment and most of our crew back to town and then we waited for the Beaver to come pick us up. The Beaver arrived and we refueled it with 80-octane aviation gas that we had in drums at our camp. Meanwhile, the pilot looked with concern at some slight damage to the right wing tip that his crew had inflicted back at Tripoli airport when they wheeled the plane out of the hanger. After refueling the plane we found our supply of aviation gas was much lower than we expected and on interrogating our crew we found out that they had been putting aviation gas in our trucks for the last few days by mistake. We then flew back to Tripoli and on the way passed over a dozen or so British army tanks on maneuvers.

Back in Tripoli we were ready for a week of field leave. We bachelors amused ourselves with snorkeling, spearfishing, sailing and dining at the excellent Italian restaurants. I took scuba lessons at The Underwater Explorers Club and then bought a used Siebe-Gorman aqualung from the owner. We also re-visited the fascinating Roman ruins at Leptis Magna and Sabratha (these two Roman cities and the ancient city of Oea, now underneath Tripoli, formed the Roman "Three Cities" or "Tri-poli").

On September 9, 1957, my fellow geologist, Jim Carter, and I left Tripoli for a vacation on the French Riviera and Spain. We flew by Douglas DC-3 to Tunis, Tunisia, and then boarded an Air France DC-6 for Marseilles, France. The plane landed and we raised a fuss because we could not find our baggage and could not speak French. Finally, the authorities made us understand that we were not in Marseilles at all but had made a stop at Ajaccio on the island of Corsica! They also told us that our plane was on the runway with engines running waiting for us to get back on board. On our arrival over the Marseilles Airport, we found ourselves in the middle of an airshow and so our pilot made a low pass over the field and then pulled up very quickly to give the crowd, and us, a thrill. We checked into the very modest Hotel Mediterranee. We had some good meals in Marseilles and visited the famous, or infamous, French Foreign Legion Fort, the Chateau D'If, at the harbor entrance. Then we rented a car and drove east along the south coast of France making stops at several famous resort towns, including Cannes and Nice. The weather was beautiful and we ate at nice restaurants, visited nightclubs and casinos, and lounged on the beaches. It was all a nice change from the Sahara Desert.

After a nice week on the Riviera, we boarded a second-class train car in Nice, France to travel overnight to Barcelona, Spain. For the equivalent of $18 we found ourselves in a sleeping compartment for six people with no curtains or sheets, just a pillow, a blanket and a bare wooden bunk each. At 6:30 a.m. the next morning we were awakened and surprised when a porter threw our pillows and blankets out the window. We had arrived at the Spanish border and had to have our passports checked. Upon arrival in Barcelona (pronounced "Barth-alona" in the Castillian Spanish spoken there) we ended up in the modest Hotel Ramblas at a cost of $37 a night including breakfast and lunch. In Barcelona we feasted on seafood including my first taste of *calamari* (slices of squid fried like onion rings), which I liked from the start. In those days a huge dinner with wine and champagne cost about $2.50 per person. We took a night tour and visited some nightclubs. Then, we flew by DC-3 to Madrid where we checked into the Hotel Paris. In Madrid we toured the city and watched the bloody bullfights at the Plaza de Toros. We also dined on roast suckling pig at the famous Casa Botin Restaurant founded in 1725.

When our stay in Madrid was over, we flew to Rome for a brief visit. Then we flew to Catania, Sicily, where our pilot kindly circled the smoldering and yellow, sulfur-stained summit of the Mount Etna volcano before landing. After a night in Catania, we flew to Tripoli where we returned home on September 30 after an interesting and enjoyable three-week vacation.

Back in the Tripoli office, I was teamed up with Jim Carter for future fieldwork. The two of us worked in the office formulating our field plans. Meanwhile, we spent all our spare time snorkeling off the rocky beaches west of Tripoli. Our underwater sightings included groupers, stingrays, moray eels, octopus, Roman pottery, and World War II bombs and grenades.

On October 13, the following letter was posted in our office: "This is to advise you that armed bandits are operating in Concession 26 northeast of Dirj. On about Wednesday of this week, approximately seven (7) men with rifles stopped two Robert H. Ray trucks and hijacked all cigarettes, tea and sugar." Apparently the bandits were nomadic tribesmen, or Bedouins, riding camels.

On October 16, Jim and I left Tripoli to make a geological reconnaissance trip of the high plateau country south of Ghadames oasis, known as the Hamada Al Hamra (barren red plateau). We flew in the Oasis C-47 with two Libyan drivers and equipment to our dirt airstrip at Dirj. Over the next couple of days, Jim and I borrowed a Dodge Powerwagon and a Land Rover from another camp and prepared for our excursion to the Hamada. We loaded our two Powerwagons with supplies and got into our Land Rover to start our trip south. We followed Wadi Amasin south until we reached Astrofix T-15, a point our company surveyor had fixed by shooting the stars. Next day we measured some geologic sections on a limestone-capped butte near T-15 and then continued south down Wadi Amasin until we reached the foot of the high cliff marking the edge of the Hamada. After some searching, we found a place where we could work our vehicles up the escarpment onto the plateau but it was a treacherous ascent up a narrow, rocky path.

Once on top of the Hamada we found it to be a flat, rocky plateau without much vegetation. We headed southeast to a spot on our aerial-photographs that we thought might be Bir Ghaziel (Arabic for "the water hole of the gazelle"). On the way, several gazelles ran along in front of our Land Rover at high speed. It reminded me of the report I had seen by an Esso geologist. With tongue-in-cheek, he reported driving along on the Hamada when "a dreaded gazelle charged into his Jeep backwards and was killed"; no doubt there was fresh meat on the table for them that night. We arrived at the spot we had seen our aerial photos and thought was Bir Ghaziel, but found it to be only a grassy depression with camel tracks leading into it, and not a well. We camped for the night under the stars. Having failed to find the well, we decided to leave the plateau and return to our base camp.

The next day we drove along the edge of the Hamada looking for a place to get back down into Wadi Amasin. It looked like the descent was going to be even more exciting than the ascent. We finally found a washed-out part of the cliff where we were able to work our vehicles slowly down to the desert floor. For several days we reconnoitered around the edge of the Hamada and looked for the Bir Amasin water well, but never found it. Finally, we returned to our camp near Dirj and waited for the Oasis C-47 from Tripoli. The plane arrived and we headed home to Tripoli. Also aboard our plane was a Libyan who had been fired by the manager of a nearby Robert H. Ray seismic camp. It seems they had sent this man over to their bar (they lived a lot better in their seismic camps than we did) to wash the glasses and dishes, but 30 minutes later they found him passed out drunk on the floor.

Jim and I arrived back in Tripoli on October 21 and spent a week in the office preparing for our next trip to the desert and getting our personal affairs in order. The weather was stormy and windy and so we could not do any snorkeling or scuba diving, but we entertained ourselves with sidewalk cafes, Italian restaurants, and our other bachelor pursuits. One day at a bar we drank numerous gin-and-tonic cocktails and this prompted us to calculate from the amount of quinine listed on the tonic bottles that it would take 35 gin-and-tonics a day to get enough quinine to prevent malaria!

10 ✳

Saharan Reconnaissance

On October 31, 1957, Del Weigand, Jim Carter and I left Tripoli, Libya with a heavily loaded convoy of trucks. Our mission was a long geological reconnaissance trip to Oasis Oil Company's unexplored Oil Concessions 59 and 71, located almost a thousand miles away in the Sahara Desert of southeastern Libya. It was winter with rainy, northwesterly winds sweeping along the Mediterranean coast. We encountered thunderstorms as we drove east from Tripoli along the coast road, a two-lane asphalt strip extending to Cairo in the east and Tunis in the west. The highway was lined with tall eucalyptus and date palm trees as we drove along Libya's coastal oasis belt of olive and citrus groves.

We passed the magnificent ruins of the ancient Roman city of Leptis Magna, which had reached it peak under Emperor Septimius Severus around 200 AD, and continued east to the town of Homs, where we stopped for lunch at a small, rustic restaurant. We continued east passing through Misurata and then stopped by the side of the road for the night. After dining on canned ham and other simple fare, we set up cots and slept in the open under cloudy skies. Fortunately it did not rain.

We woke up the next morning to find our sleeping bags covered with heavy dew. Our Libyan cook fixed breakfast. While cleaning up the campsite prior to departure, we found an old British army shoulder patch and some tin cans dated 1940. Apparently, some British soldiers had camped on this same spot during the 1940 to 1943 North African campaign of World War II. We headed off down the coast road, which now turned southeasterly following the curving coastline of the Gulf of Sirte. Moving southward in Libya meant going towards the desert and the terrain started becoming sandy and dry. We drove through the town of Sirte and ended the day at Nofilia. Here, waiting to join us was a field crew specialized in detecting and destroying World War II landmines, which number in the millions all over northern Libya. In 1942, General Field Marshall Erwin Rommel, commander of the German Afrika Korps, made a last-ditch stand in this area against the advancing British General Montgomery by establishing his "El Agheila Line." The defenses consisted mostly of minefields because of lack of high ground for defensive positions.

Our convoy now consisted of a dozen vehicles including Land Rovers and heavily loaded trucks. Personnel consisted of three geologists (Weigand, Carter and I), two mine-clearance specialists, a British mechanic, an Italian truck-driver, and about 30 Libyan drivers, mine-sweepers, cooks and helpers. Mine clearance was, needless to say, very important to us and was carried out by Joe Sainato, a former U.S. Army munitions specialist, and Carl Brandt, a former German army officer who had helped Rommel lay the land-mines in Libya. This was certainly the largest geological party I had worked with.

Our convoy continued east along the asphalt coast road until we came to what the British in WWII called "Marble Arch," a high marble monument arching over the road erected by the Italians in 1937 to commemorate the completion of the coast road. They named it The Arch of the Fileni and it marked the border between the Libyan provinces of Tripolitania and Cyrenaica. Legend has it that in ancient times the two Fileni brothers running from the Greek City of Cyrene to the east met runners from Carthage in the west to establish the border between their two lands at this point.

We experienced a considerable delay passing through Marble Arch due to customs red tape at the police posts on both sides of the provincial border. Continuing east we came to El Agheila, famous as the north end of Rommel's El Agheila Line, and then on to Agedabia. Agedabia is a rather dreary place that was used as a fortress by both the German and British armies as they fought back and forth along the coast road; it was still surrounded by rings of mine fields. The town now housed a Libyan insane asylum, as was well known to oil geologists; we referred to going crazy in the desert as "being fitted with an Agedabia sport-coat (straight jacket)."

At Agedabia, we left the well-paved coast road and drove south along a well-marked desert track amidst scattered minefields. Although the desert war between Rommel and Montgomery had been over for almost 15 years, the land mines were well preserved by the dry desert climate and still quite lethal; several Libyan Bedouins and their camels were killed each year by the mines. The mine clearance crew swept an area clear of mines and we camped for the night. A Welshman from a nearby geophysical (gravity survey) crew joined our party to guide us south into the desert to his camp.

The next morning our convoy headed south. The going was rocky but good for about 45 miles and then we ran into a sea of pure quartz sand. Our heavy trucks immediately started getting stuck. The Kenworth 18-wheeler was especially bad as its four-wheel drive stalled the engine every time it got stuck and we had to pull it out with two Ford Power Wagons. We spent all afternoon digging sand out from under wheels with our bare hands and shovels, and shoving steel sand tracks under the wheels. We finally made about 50 miles and camped for the night. The desert here was very flat and featureless with not even a blade of grass in sight. As we bedded down for the night we discovered large pieces of petrified wood scattered around on the ground; millions of years ago this part of the world had been a swampy area.

We continued driving south and the trucks immediately started getting stuck again. It became obvious that the trucks were so heavily loaded that they broke through the crust on the top of the sand and then got stuck. By this time, we were out of the minefields so Jim Carter and I went ahead in our Land Rover and found a better path on slightly higher and firmer ground. We pulled all the trucks out of the sand and slowly moved them forward onto the higher ground. We finally came to a hard, flat, sandy surface and the convoy made much better time.

The next morning was November 5, 1957, and my 26th birthday; but, there was no birthday party for me. We awoke to find that a heavy fog had settled in. Jim and I drove ahead in the fog to scout out the route. The fog lifted a few hours later and the convoy drove ahead with a good, hard, sandy surface. In the afternoon we drove around an area of high sand dunes and arrived at the camp of Robert H. Ray Gravity Party No. G-138, where we were greeted by the party chief, George Sweat. Dinner that night was a real treat for us with drinks, hors d'oeuvres, fried chicken and ice cream.

Our camp was about 150 miles south into the Sahara Desert in Oasis Oil Company Concession Number 59. The area was completely desolate, except for a small oasis (Gialo) marked on our maps some 80 miles to the east. Jim Carter and I left the camp on a reconnaissance trip and headed west. After driving about 30 miles we found ourselves at the top of a 500-foot-high cliff or escarpment. Beyond the escarpment to the west lay a sea of high sand dunes stretching to the horizon. We returned to camp to plan our geological surveys. On return to camp, Joe Sainato, the mine-clearance chief, told us he had scouted to the south and in a dry river bed, or *wadi*, leading down through the escarpment, his crew found several World War II trucks that had been bombed, and several large, unexploded bombs. Markings on the trucks and bombs indicated that this had been a Canadian Army unit bombed by German planes.

According to our maps there was a small water hole at the base of the escarpment known as Bir Zelten. The next day, Joe Sainato took his crew to check out the water hole. He came back to report that on the way to the water hole he had found a small grove of bushes with a desert track making a detour around it. After sweeping the area with mine detectors, Joe found that German troops had planted Teller anti-tank mines among the bushes. The Germans probably figured that British tanks or trucks coming down the track would take the short-cut through the bushes, or think that the water hole was located there, and be blown up by the mines. Joe found that a Robert H. Ray vehicle had just missed one of the mines by a few feet.

We loaded up our convoy of trucks and left the Robert H. Ray camp. After traveling 15 miles west towards the escarpment, we stopped and set up a permanent camp for our surveys. Our location according to our military maps was latitude 28 degrees 20 minutes north by longitude 20 degrees 19 minutes east. Our supplies were getting low and the Oasis C-47 cargo plane was scheduled to land supplies for us in a few days at the nearby Robert H. Ray Camp. Del Weigand, one of the geologists, got on the short-wave radio and read off a list of supplies to our base in Tripoli.

Weigand wanted them to send us some beer but he didn't want to mention the word "beer" on the radio because all the desert camps listened on the same frequency and beer was not allowed in some camps. So, he requested "four cases of Coor's mountain valley sparkling water" and figured our American Chief Geologist, John Goffe, would figure out that we were using a play on a Coor's advertisement and really wanted beer. When the plane arrived several days later we were very disappointed to find they had sent us four cases of Italian mineral water! It seems John Goffe was out of town and our order went to our British warehouse supervisor, Eric Johnson, who did not have a clue what we really wanted. The "Coors sparkling water" incident became a legend in the Libyan Desert and is even recorded in the book, *Portrait in Oil, A History of Marathon Oil Company*, written by Hartzell Spence for Marathon Oil Company in 1962. However, Spence used a lot of author's license in re-telling the story, including leaving me out.

To get our geological surveys underway, I went to the nearby Robert H. Ray camp and plotted their surveyed gravity stations onto our aerial photographs. As I drove around the area I saw a number of gazelles and they would often run in front of my Land Rover at full speed for some minutes before veering off to the sides.

George Sweat came over from the Ray Camp one day and told us he had been scouting a route west through the sand dunes for his next camp move. He said he was driving just south of Bir Zelten when he came over a hill and was shocked to find his vehicle surrounded by 90 British Mark II and French "waffle" anti-tank mines exposed on the surface. He carefully backed out over his own tracks and returned to camp. The next day the mine clearance crew went over and exploded all the mines in place.

Del Weigand and I drove around the area and examined the rock outcrops. On November 12, we came to the *wadi* with the bombed Canadian trucks. We found a couple of Ford trucks blown to pieces, several unexploded bombs, and the whole area littered with helmets, machine-gun bullets, holsters, belts, gasoline cans, and other items.

The Oasis C-47 landed near our camp and the Oasis radioman, an Italian named Trotolo, installed a large Viking II radio transmitter and a National NC-183 short-wave radio receiver, so our radio communications would be more reliable at our location some 600 miles southeast of our headquarters in Tripoli. I was delighted to see this really nice amateur radio type equipment, which I was very familiar with from my amateur radio days back in the States.

On November 14, the Oasis C-47 landed by our camp and the pilot took me up for a two-hour reconnaissance flight over the area that we were surveying, which gave me a much better idea of the lay of the land and where the rock outcrops were located. The plane landed at camp and most of our party got on board for the flight home to Tripoli after 12 days in the desert. After five days of leave and office work in Tripoli, we flew back to our camp near Bir Zelten in Concession 59. After we landed, another C-47 arrived and it was quite a sight to see two C-47s parked next to each other out in the middle of the desert. Robert H. Ray Geophysical Company owned the other plane and on board was none other than Robert H. Ray himself making an inspection trip of his Libyan camps.

The next day, Jim Carter and I got our geological gear in order and ran a few short traverses to check our plane table and alidade surveying instruments. We used this equipment to make geological maps by: placing the alidade on a base map taped to a board on a tripod; taking a bearing on a surveying rod placed on a rock outcrop some distance away; reading the distance using two parallel cross-hairs in the alidade's telescope, drawing a line on the map along the ruler-like base of the alidade, and marking off the distance. Notes on the alidade readings were kept by the "instrument-man," in this case me, and notes on the geology of the rock outcrops were kept by the geologist, in this case Jim Carter.

In the desert the biggest problem with reading an alidade was heatwaves. The wavy, blurry atmosphere distorted the view of the rod in the telescope making it very difficult or impossible to accurately read the distance from the tiny image of the rod located a mile or two away. Because of this problem, our work was usually limited to early morning and late afternoon. In addition, as previously mentioned, our stations were so far apart that I had to make a correction for the curvature of the earth.

On November 22, 1957, Jim Carter and I started a geological survey loop along the high escarpment near our camp. We chose to follow the escarpment because it revealed the best rock outcrops in the flat featureless terrain. The weather was very nice with warm, but not hot, sunny days and rather cold nights. One of the German mine-clearance men checked out the *wadis* in the area where we were working and blew up 19 handgrenades and several smoke bombs at one site near us. He also found an Italian helmet, a British helmet with a bullet-hole in it, and some partially burned war maps.

WORLD WAR II MILITARY ARTIFACTS.
Near our camp in Concession 71, our mine clearance crew found
these items. Seen are a British helmet, gasmask, bomb fins, a French
"waffle" land mine, and the top of a German Teller anti-tank mine.
I am wearing the helmet and had a mustache in those days.

As Jim Carter and I made our alidade survey, Del Weigand checked out rock outcrops to the south around the escarpment. One day he came back to camp to report that he had found a whole graveyard of large, fossilized, marine, animal bones lying in the sand at the foot of the escarpment. He brought a large skull and collarbone back to camp to show us. After a few days of rest, Jim Carter and I started a plane table and alidade traverse along the northwestern extension of the Bir Zelten escarpment. One day we drove to a point on the escarpment right above the Bir Zelten water hole. We could see camel and vehicle tracks converging on a grove of desert bushes, which marked the water hole. Around the hole lay a large number of petrified trees lying in broken sections on the sand.

On our way back to camp a desert fox, known locally as a *deeb*, with a bushy tail and large ears, started running in front of our Land Rover. He ran ahead at almost 35 miles per hour for some time and then suddenly darted off to the side and disappeared. I was reminded of the story that a Libyan had told me about the desert fox as we sat around our campfire in the desert. According to him, if a person falls asleep at night on the open sand, a desert fox will creep up to you, urinate on its tail, flick the urine in your face to make sure you are asleep, dig the sand away from under your head so your head tilts back, and then grab you by the throat and kill you!

Jim Carter, Don DePriest and I returned to the escarpment at Bir Zelten a few days later. We carefully climbed down the almost vertical slope to measure and describe the Miocene-aged rock layers exposed in the cliff, which was about 500 feet high. At the foot of the cliff we found a veritable forest of petrified trees. The palm-type trees were excellently preserved as dark red, petrified wood. Also scattered around were the fossils of very large vertebrate animals. Whole skeletons of skulls, teeth, vertebrae and other bones lay on the sand. The vertebrae were the size of washtubs and were probably the fossilized remain of whales that had once swum in an ancient ocean. So, we were looking at evidence that had washed out of the layers of this cliff that this area had once been covered by a great ocean and then later been uplifted to become a swampy area of palm trees. As we surveyed the rock layers and fossils, we were attacked by swarms of flies from the nearby water hole.

The next day the C-47 landed with supplies and radio repair parts, which I installed. We then started preparing the trucks, equipment and supplies for a long reconnaissance trip south into Concession 71, which had been newly awarded to Oasis Oil Company.

On December fifth, our convoy left camp for Concession 71, but we soon fell victim to numerous delays. First, we had to send a Land Rover back for forgotten maps and survey equipment. Then our 250-gallon water trailer broke down while descending the Bir Zelten escarpment; it was towed back to camp and replaced by our Ford Tandem truck with two 500-gallon water tanks. Then our Libyan cook discovered he had brought *cous-cous* semolina instead of sugar, so another Land Rover was sent back to camp. Then Don DePriest remembered he had forgotten his medicine, so he went back to camp. We only made it 35 miles south that first day. The next day we traveled south deeper into the desert on a flat, firm sand surface at speeds of up to 40 miles per hour. The terrain was entirely featureless, with not so much as a bush in sight. Using old military maps, we followed a desert track, consisting of vehicle tracks in the sand with occasional metal markers sticking out of the sand, from Bir Zelten south towards the Oasis of Tazerbo.

**THE CREW TAKES A SOCCER BREAK ON
THE WAY TO TAZERBO OASIS.**
My fellow geologist, Jim Carter, is on the left. Seen are three
Landrovers complete with surveying equipment and canvas water
bags. We also had a water truck and mine clearance personnel.

To aid in our navigation while driving, Carl Brandt, our German mine-clearance supervisor, used a "sun compass" installed on the fender of his Land Rover. This instrument was devised by the British Long Range Desert Group during World War II and consists of a horizontal disk marked off in radial sections with a vertical sundial rod mounted in the middle. By watching the shadow of the rod on the disk, and correcting for the time of day, the sun-compass helps to steer a course across a featureless desert without frequent stops to take compass readings.

About noon the second day out, we came across a World War II British truck with the engine and wheels gone. Carl Brandt went to investigate and reported back to us that the bed of the truck was filled with boxes of handgrenades, and scattered all around were landmines, fuses, machine-gun bullets and other munitions. We carefully drove around the area and continued south until we arrived at Tazerbo Oasis at 5 p.m. and camped at the Police Post for the night. We had covered about 200 miles during the day.

The next day we took three Land Rovers and a Dodge Power-Wagon out to explore the eastern end of Concession 71. We drove to the eastern border of the concession near the water well at Bir-el-Atasi at longitude 22 degrees east. Then, we swung south to the Oasis of Bzema, traveling at speeds up to 55 miles per hour over the flat, sandy plain. Bzema consisted of several high buttes and mesas of very dark, almost black, rocks sticking up out of the sand dunes with an oasis of palm-trees and small lakes wrapped around the south end. We stopped to examine and sample the rock layers in a 500-foot high cliff. Bzema is about 500 miles south of the Mediterranean Sea and the farthest south in the Sahara Desert that I ever reached. We turned around and as we drove northwest back to our camp at Tazerbo we encountered several long lines of very high sand dunes crossing our path. Going over these dunes was tricky because we had to keep up our speed to keep from sinking into the sand and getting stuck. So, we picked the lowest part of each line of dunes and went over it as fast as we could go. This resulted in several thrills as we often found ourselves sailing through the air over the steep backside of a dune.

On one occasion, the first Land Rover in our convoy sailed over the back of a large sand dune and got stuck nose-down in the sand.

Jim Carter and I were second over and just missed the stuck vehicle as we got stuck ourselves. We quickly ran clear as the third Land Rover sailed over the dune and also got stuck; Don DePriest and Del Weigand jumped out and ran as fast as they could because they knew the Dodge Power Wagon was coming over right behind them. Fortunately, the Power Wagon truck slowed down as it climbed the dune, stopped as it reached the top, and then slowly descended down the back. We continued northwest back to our camp at Tazerbo without further incident.

That evening as we sat around our campfire we were surprised when a European wearing a French beret walked up and greeted us. He told us he was Professor W.W. Rajkowski from the University of Durham, England. He said he was Polish, but now a British subject. He was delighted to share our western-style dinner with us and told us his story. Rajkowski said he had been wandering in the desert for over nine months with a Libyan guide and two camels, which were now staying at the Police post. As an anthropologist, he had been far to the south studying the local inhabitants in the Tibesti Mountains and Kufra Oasis. He was presently living at the Tazerbo Police Post and preparing to travel west across the desert to the Oasis of Sebha.

The next morning, Professor Rajkowski joined us for breakfast and then took Del Weigand, Jim Carter and me along with him to find "the castle of the last Tubbu King of Tazerbo," which he had heard about but not yet seen. Apparently, these mysterious Tubbu people were people from Central Africa that had moved into the Sahara. They constantly fought with the Tuaregs, who finally drove them back south. We walked through Tazerbo Oasis guided by Rajkowski and local residents, and then came to ruins of a mud-brick citadel surrounded by ruins of circular animal pens. This "castle" was small and in a very poor state. Rajkowski estimated that the Tubbu people had abandoned it about 150 years ago. Following this tour, Rajkowski left our camp and we never saw him again.

We broke camp at Tazerbo and our convoy traveled northwest along the desert track towards Zella. About mid-day, we came across a World War II British army truck with a complete shower unit on it, a burned-out British truck, and a World War II airstrip marked by gasoline drums. That day we crossed about 140 miles of flat, sandy, gravel plain, known as a *serir,* with a few sand dunes before we stopped to camp for the night. The next day the trucks and one Land Rover continued north to join the tracks we had made when we traveled from our Bir Zelten camp to Tazerbo. Using two Land Rovers, Don DePriest, Del Weigand and I drove northwest to check out some hills shown on our maps. In the desert, hills usually mean rock outcrops which provide the only geological data available in sand covered areas.

After traveling some 35 miles northwest, we found some low buttes capped with hard limestone located near the western border of Concession 71. After examining the outcrops we headed due east to re-join our convoy. On the way, we encountered more parallel ridges of sand dunes and got several more thrills as we drove over the dunes at high speed to find a very steep drop-off on the other side. We got stuck in the sand several times but got out using steel sand-tracks, which we always carried (another British Long Range Desert Group invention from World War II). We rejoined our convoy and followed our old tracks back to our permanent camp near Bir Zelten in Concession 59. We had covered a total of about 900 miles on our reconnaissance trip to Concession 71. It was now December 11, and the winter weather in the desert consisted of clear, sunny days with a cool breeze, followed by cold nights around 35 degrees Fahrenheit.

Back in camp, equipment was repaired and stowed in preparation for our flight back to Tripoli. The last flight of the year, known as the "Santa Claus Special," was scheduled to pick up all American personnel in the desert and take them to Tripoli for the Christmas holidays. The Oasis C-47 arrived and flew us back to Tripoli. I spent the Christmas holidays with the other bachelors. Our activities included: skin-diving and spear-fishing; going to movies, and attending Christmas parties put on by the senior managers of Oasis Oil Company, including Resident Manager Albert F. Lager and his wife Martha, and Chief Geologist V. L. "Jack" Frost and his wife, "Lou." On Christmas Day, Jim Carter and I donned our newly purchased Italian wetsuits and went spearfishing five miles west of Tripoli. The water was very cold and clear, and we found the sea floor scattered with fragments of Roman pottery. The wet suits were nice and warm once the initial surge of cold water down our backs had warmed-up from our body heat.

On December 27, our exploration manager, Jack Frost, assembled all the geologists and informed us that the Oasis geological and geophysical budget for 1958 had been greatly increased over 1957. This meant that we had a very busy year ahead. A few days later, Tripoli was buzzing with the exciting news that Esso Oil Company had struck oil at their Atshan No. 3 well in Fezzan Province, deep in the western Libyan Desert. Rumors had oil flowing to the surface at 1,200 barrels per day from a geological formation at a depth of 2,100-2,140 feet. Although there had been several showings of oil in Libya up to this date, this appeared to be the first commercial strike. We were all very excited and encouraged by the news.

On New Year's Eve 1957, we bachelors all went to a New Year's Eve party at John Dunlop's villa. We got back to our bachelor apartments in downtown Tripoli at three o'clock in the morning and got an idea that was to create a legend among the expatriates of Libya. Jack Redmond, Jim Carter and I rigged up a 35-mm. slide projector and pointed it out the second floor window onto the side of the building across the street. We then projected slides of semi-nude women across the street. The building across the street was so far away that we ended up with projected nude photos almost thirty feet square. As we showed the slides, we started hearing cars in the street below screeching to a stop and parking under our window. Then, a British army truck full of troops heading back to their barracks pulled over in the street below us. Every time we changed to a new slide a big cheer went up in the street!

On January 2, 1958, I boarded the Oasis C-47 and flew back to our camp (Camp 5ALQ - our radio call letters) in Concession 59 with Jim Carter, Del Weigand and Don DePriest. The nights were still cold and I used an Aladdin kerosene heater to heat my tent to keep warm. One evening after dinner, I returned to my tent to find that the heater had gone haywire and covered the roof of my tent with two inches of black soot.

Jim Carter and I resumed our alidade survey around the escarpment edge of a diamond-shaped plateau extending northwest from Bir Zelten. On the way to the field one day, twelve gazelles ran straight ahead in front of our vehicle at full speed for quite a while. It was strange the way the gazelle often did this, because they could have run away from us in any direction they wanted to. After seven days, Jim and I worked our way back to where we had started our survey loop. We had carried our elevation and bearings in a loop for almost 100 miles without any place to check them along the way. This was the moment of truth for me as the instrument-man; we would now see how accurately I had recorded my alidade shots and drawn my map. I drew the last line on our map and was pleased to see that the final location plotted within 75 feet of the point where we had started. I then sat down to compare the final elevation with the elevation we had started with. I was amazed to report to Jim that after almost 100 miles of uncontrolled loop we had tied the starting elevation within two feet. Jim's immediate words were "don't recheck your calculations, we are done." On the way back to camp another desert fox briefly tried to outrun our Land Rover. Our work on the Bir Zelten escarpment of Concession 59 was now finished. Our geological mapping indicated a strong anticlinal axis running north out of our concession, which indicated good oil potential. Several years later, Esso discovered their huge Zelten oil field on the northern extension of the axis we had mapped, and Oasis found several oilfields on the southern extension.

It was now time to move west to a camp that Weigand and DePriest had set up about 120 miles west of the Bir Zelten escarpment. The new camp was near the relocated Robert H. Ray gravity survey camp in the western end of Concession 59. Jim and I drove west all day to the new camp. On the way we crossed a 30-mile-wide stretch of sand dunes, but we were able to find our way through without getting stuck. That night we had dinner at the Robert H. Ray camp, which was a treat as they had a much larger camp and ate much better than we did. After dinner we sat under the stars and watched a movie. There was some excitement when it came time to change a reel and the operator feeling for an empty film cans in the dark discovered the camp dog had thrown-up in it.

The next day a convoy of landmine clearance vehicles arrived. And, International Aeradio Ltd. (IAL), our radio service in Tripoli, advised us that a United Geophysical Company Land Rover carrying an American surveyor and two Libyans was lost in the desert. They also sent us the following two radio messages on January 13:

<u>Message No. 3381 from 5ALL to 5ALQ:</u> For identification purposes if your transport is sighted by aircraft, stop and reverse 100 yards. The aircraft will then ignore you and continue.

<u>Message No. 3384 from 5ALL to 5ALQ:</u> What stocks of aviation fuel have you and what is its octane value? Your assistance in supplying petrol to Royal Air Force (RAF) aircraft in search may be requested. If petrol available please give airstrip co-ordinates.

Joe Sainato, our Mine Clearance Supervisor, arrived on the Robert H. Ray C-47, together with Carl Brandt, back from leave in Benghazi. Sainato filled us in on the latest Libyan news. He said some Libyan was driving around the eastern coast road in a stolen Mobil Oil Land Rover with a load of handgrenades. He threw a grenade into a Geophysical Service Corporation (GSI) camp but the explosion did not injure anyone. Then he stopped a gasoline truck south of Agedabia and threatened to blow it up with seven handgrenades hung around his neck unless they gave him the gasoline, which they did. Sainato also told us the lost United Geophysical Land Rover had not been found. Carl Brandt related that a brand new kitchen trailer just off a ship in Tripoli had blown up in flames while on convoy to the desert because someone had left the butane stove turned on.

Jim Carter and I started a new plane-table traverse to study the geology just west of our camp in western Concession 59. The weather was clear, calm and warm. One day we noticed some unusual blue-white slabs of rock showing through the sand. We struck one with a rock hammer and were surprised to find we had broken into a cavity lined with long, delicate, light blue crystals. The whole area was scattered with these crystal geodes and so we photographed them and collected specimens. [The crystals were later identified as a strontium mineral called celestite. I prepared a report on the celestite occurrence for the Government of Libya and donated a slab of crystals to the museum inside the Old Tripoli Fort.]

Oasis informed us on the radio that the lost United Geophysical Land Rover had finally been found. When the three men in the vehicle found they were lost and ran out of gas, they did the wrong thing and left their vehicle (standard emergency instructions in the desert were to stay with your vehicle and set a tire on fire when aircraft appeared in the area). The three men wandered around in the desert in all different directions with only beer and oranges to live on. Before they were found they had worn out their shoes and the two Libyans had been forced to drink beer to stay alive even though their Moslem religion forbade it. The three finally straggled into a GSI camp after eleven days of wandering lost in the desert. They were lucky to be alive.

The United Geophysical personnel were found, but now we were advised by radio that two Gulf Oil Company geologists were lost in a sand sea south of Concession 26 on the Algerian border. It is easy to get lost in the featureless desert as there are no land marks to use and distances are very difficult to judge. I thought I saw a truck in the distance one day and drove towards it only to find that it was a rusty one-gallon gasoline can. I never got lost but I came close to it a couple of times and the feeling is quite scary, especially when you are alone with no radio and darkness is coming on.

After two weeks in the desert, Jim Carter and I flew back to Tripoli on leave. At our bachelor quarters, we found that Bill Noonan had brought a baby gazelle in from the field. The gazelle, which he had named "Ace," had trouble walking on the polished marble floors so Bill made leather shoes for it to wear.

On January 20, 1958, I went to see a Ministry of Communications Radio Engineer, Mr. Faturi. He examined my American amateur radio license and commercial radio operator's license, and said a Libyan amateur radio license would be issued to me immediately without an examination. This meant I could now take amateur radio, frequency crystals with me to the desert, install them in the company transmitters, and talk to other "ham" radio stations all over the world. I soon received my Libyan license to operate as amateur radio station 5A4TV and talked to ham radio stations around the world from our desert camps.

One day as I was sitting at a sidewalk café in Tripoli reading a copy of the *Cyrenaica Weekly News*, I was shocked to see an article entitled, "Desert Tragedy." This article said a Professor Rajkowski had died in the Haroush District of the Libyan Desert and that his guide had come into Benghazi to report it. I immediately went to see Lieutenant

Mukhtar, the police liaison officer with Oasis Oil, and told him that we had seen Rajkowski alive in Tazerbo Oasis on December 8, 1957. I gave him my photograph of Rajkowski taken at the time, and it was concluded that we had been the last westerners to see him alive. I wrote out a detailed report for the police.

26th January, 1958   SUNDAY GHIBLI

Researchman Feared Dead In Desert

Lone Scientist Is Weeks Overdue At Mail Point: Unknown Body May Be He

WHAT has happened to Mr W.W. Rajkowski, lone desert traveller and scientist, who was last seen in the Western Desert at the end of November?

Some months later, I sent a copy of my police report to the University of Durham in England to tell them about our meeting with Professor Rajkowski in the desert. The chairman of the Geography Department wrote back to thank me and said that they had received very little information from the Libyan police. He said he would pass along my information and expression of sympathy to Mrs. Rajkowski.

On January 23, Jim Carter and I flew back to our camp in western Concession 59. On the way to our camp, the C-47 overshot the Oasis camp in Concession 32, where we usually stopped to refuel, and turned back towards the coast to get its bearings. After flying over Marble Arch, we finally found the Concession 32 camp and then flew on to our camp in Concession 59. Jim and I continued our geological survey in western Concession 59. One day we made a reconnaissance trip south some ten miles outside our concession and found a black, volcanic mountain, or volcanic plug, several hundred feet high sticking up out of the flat, sandy desert. Two walls, or dikes, of black basaltic rock radiated out into the desert from the volcanic plug. According to our military maps, this black volcanic mountain was known in Arabic as Gleb-el-Barrut, which our Libyan crew said meant, "boiling teapot."

Carl Brandt came back into camp from Zella Oasis, where he had heard about the death of Professor Rajkowski. He said Rajkowski's

Libyan guide had stopped in Zella on his way to Benghazi, and while there he told the police that Rajkowski had died of thirst, but he had survived by slaughtering a camel and drinking the fluids inside. We all wondered if this was the true story, or whether his guide had killed Rajkowski for his camels and other belongings.

As February arrived, we continued our geological surveys in western Concession 59. Del Weigand and Don DePriest flew in from Tripoli to help so we could double our mapping efforts. The weather was very nice, but beginning to warm up. Daytime temperatures were now reaching 85 degrees, but the nights were still cold. While listening to my short-wave radio on February 1, I heard that the United States had launched its first earth satellite into orbit. Named "Explorer I", it was launched using a Jupiter-C rocket. On February 5, Jim and I completed our plane-table geological survey in western Concession 59. I was happy to find that we had tied our survey loop elevation to within less than a foot and our horizontal error was less than 300 feet. This was amazing accuracy for such uncontrolled surveying.

Our next project was to survey a large triangle of accurate survey points in the western end of Concession 59 to be used later for more detailed geological and geophysical surveys. On February 6, Pierre Vuille, our Swiss surveyor, flew to our camp to locate several "astrofixes," or survey points from sighting on stars. These precisely located points would tie our geologic surveys into the Oasis base maps. We started laying out a triangulation system with our plane table and alidade to incorporate the new astrofixes. One morning we came to a small cliff and down below was a family of Uaddan antelope with long curving horns. These were pretty rare in the desert, so we stopped to watch them. They finally spotted us, stood and watched us, and then trotted away.

Pierre Vuille's survey crew started causing labor problems in our mine clearance camp supervised by Carl Brandt. The problem was that Vuille paid his men more and supplied them with bottled mineral water for their meals. Brandt's men demanded equal treatment, but Brandt refused to give in to them. I spend an entire night with Pierre Vuille to watch him establish Astrofix No. C-36. The procedure involved: backing two vehicles together as a wind-break; using a short-wave radio for BBC time signals; taking sightings with a theodolite on seven stars (the North Star twice) in all four sectors of the sky; taking the exact temperature; and placing a white oil drum with "C-36" painted on the sides. The next morning Pierre and I spent many hours doing the intricate calculations to arrive at the precise latitude and longitude of Astrofix C-36. After doing all this, Pierre turned to me with a serious face and said, "The procedure is long, but the most important thing is to locate the correct stars!"

On February 16 we completed our geologic survey of western Concession 59 and packed up our gear for the trip back to Tripoli. Our convoy slowly made its way through Marada, El Agheila, Marble Arch, Sirte, and Misurata until finally arriving back in Tripoli.

On February 25, 1958 I was called into a meeting with Jack Frost, John Goffe and Joe Wilson, the heads of the Geology Department. They asked me about my experience in "sitting" on drilling oil wells. I told them all about my wellsite work in the Athens Field of Louisiana and the "Golden Trend" area of Southern Oklahoma. A few hours later, I was told that I had been assigned to the newly formed Subsurface Geology Department to prepare for Oasis Oil Company's first drilling rig that would be arriving in Libya soon.

Thus ended my eight months of adventurous living in tents, drinking tea with Bedouins, and exploring around the Sahara Desert of Libya. I like to refer to that phase of my life as my "Lawrence of Arabia Period."

11 ✳

Oil Rigs and Sand Dunes

In February 1958, Oasis Oil Company had assigned me to the newly formed Subsurface Geological Department. "Subsurface," which means "under the ground," meant we were to be the oil well drilling exploration department as opposed to the "surface" mapping exploration department.

For the next two months our new department planned the drilling of the company's first exploratory well in Libya. We analyzed millions of dollars worth of data and finally drew an "X" for the first place to drill on our vast 97,000 square miles of petroleum concessions (an area equivalent in size to the state of Wyoming). Next, we established geological procedures for the well and made out a checklist of well logging equipment and supplies. We planned everything as carefully as we could, but, with all the crazy things we had already experienced in Libya, we kept in mind the old Arab proverb that, "The camel driver has a plan, but the camel also has a plan."

The first well location was agreed upon and the well was given the name "A1-32 Bahi." "Bahi" means good or favorable in Arabic, and "A1-32" meant it was the first well in our Concession 32. Our Swiss surveyor was sent out in the desert to survey in the exact location by establishing an "astrofix" from the stars. Water was essential for our drilling operations, but surface water was non-existent in our area of the Sahara Desert. Fortunately, there is plenty of water under the Saharan sands left over from thousands of years ago when the climate was more temperate. A water well drilling crew was sent out to drill for water at the location and an ample supply was soon established.

Then, Tripoli's only English-language newspaper, *The Sunday Ghibli* (a *ghibli* is a sandstorm) printed a large headline saying, "Oasis Rig Convoy Heads South." Our drilling rig had arrived in Tripoli from the States by ship and was now being trucked to the drilling location in the desert of central Libya about 80 miles south of the Mediterranean Sea.

SUNDAY GHIBLI   **23rd February, 1958**

OILMEN EYE NEW SITES

Oasis Rig Convoy Heads South : Four Companies Plan Wildcat Spud-Ins

---

When everything was ready at the location to start drilling, or "spud" the well, another geologist, Jerry Hayes, and I flew out to the camp as Oasis Oil Company's first two wellsite geologists. We felt a heavy responsibility on our heads as millions of dollars of geological and geophysical surveys and years of work had gone into picking this location. If we missed an indication of oil and gas while the well was drilling it would be a very costly mistake.

Other than the drilling rig itself, the main part of our camp consisted of 12 brand new, air-conditioned, Spartan aluminum trailers. One trailer was a combination kitchen-mess hall, another was full of freezers (full of steaks and other meat frozen in the States), one a drilling office, and one was at the rig filled with our geological equipment. Most of the others were living quarters for the foreign staff and crew. We always had a doctor staying full-time at the wellsite for illnesses and injuries and he had his own hospital trailer. The doctors were mostly British and we always wondered what circumstances had prompted them to take a job in the Sahara Desert. No doubt the money was a big factor but there must have been more to it than that. One day our doctor came into his office and was shocked to find his Libyan assistant treating Libyan workers for some ills. He gave his assistant a lot of grief for practicing medicine. Then, he discovered that his assistant had broken his mercury thermometer, but had patched it up with adhesive tape and continued to use it on his Libyan patients.

Jerry and I each had a bed in one end of a trailer and our French, electrical well logging engineer, an employee of the Schlumberger Company, had the bedroom in the other end. In between our bedrooms were an office and a bathroom. This luxurious, wood-paneled, air-conditioned trailer was quite a change from the canvas tents I had been living in while doing surface geological surveys in the desert. Some distance away from our Spartan trailers were several rows of prefabricated wooden huts without air-conditioning that housed the Libyan workers.

GEOLOGIST ON OUR FIRST DRILLING WELL IN LIBYA.
I had the privilege of working on our first well in Libya, the A1-32 Bahi, in Concession 32. While I was in charge we struck oil. This was against all odds as Oasis Oil had over 64 million acres of concessions in which to choose where to drill!

Finally, on March 26, 1958, the well started to drill ("spudded") and we were off to look for oil. A large motor on the rig floor rotated the steel drillpipe that made the drill bit at the bottom of the pipe grind up the rock and make new hole. The rock cuttings were then brought to the surface by the thick, brown drilling mud, which was pumped down the drillpipe and then circulated back up to the surface to lubricate and cool the drill bit. A vibrating screen, called a "shale shaker," sifted the rock cuttings out of the mud coming up from the bottom of the hole, and a member of the crew took regular samples of these cuttings for us geologists to examine for oil and gas.

Our job was to describe the rock cuttings as they were flushed to the surface by the drilling mud. Our duties consisted of: logging the type of rocks (sandstone, limestone or shale); examining the cuttings for traces of oil using a microscope to look for brown oil staining; applying an ultraviolet (UV) light to check for oil fluorescence; and using carbontetrachloride solution to dissolve any oil stains and make them fluoresce bright yellow under the UV light. At the same time, we kept our eye on our gas detector chart that monitored the drilling mud for indications of natural gas that was being released from the rock cuttings into the drilling mud. We also gave the driller instructions to notify us immediately if the drilling bit suddenly hit a zone of fast drilling, which could indicate soft and porous rocks that could contain oil and/or gas.

As the weeks went by, I put in my 12-hour shift every day in the geology trailer monitoring the drilling of the well. While at work I listened to the BBC and Voice of America on our short-wave radio. One of my favorite BBC shows was music requests from listeners all over the world and I laughed one night when someone else in the Libyan Desert requested that they play "Love letters in the sand." During my long all-night shifts, I also learned to drink coffee for the first time. In the evenings I enjoyed walking out on the nearby sand dunes to watch the sunset; I photographed this scene many times during my desert years.

Our office trailer had a large room full of short-wave radio equipment, which was important for communication with our Tripoli office and for emergencies. As I had my Libyan amateur radio license, I was able to use our equipment at the rig to talk to ham radio operators

THE GEOLOGICAL TRAILER ON THE A1-32 WELL.
I am examining rock samples under a microscope. Also seen
(from left to right) are: an ultraviolet light box, hot plate to dry
samples, typewriter, short-wave radio and gas detector.

all over the world and they were delighted to add a contact in Libya to
their log books.

Every few weeks we would stop the drilling and our French logging
engineer would run an electrical survey of the new hole to see if any oil
or gas zones were present. This was also a test of our sample and gas
detector logging, as we should have picked up any oil or gas zones as
we drilled through them.

Our company C-47 airplane made several trips a week from Tripoli's
Idris Airport out to the A1-32 well site to deliver drilling supplies, food,
and mail; and to shuffle crew members back and forth to town. When
necessary, the plane would make unscheduled flights to deliver drilling
equipment or to meet medical emergencies. The crew on our rig was a
very mixed bag of nationalities. Americans were in charge but we also
had German, French, and Italian staff as well as a large Libyan labor
force. As a result we talked in a mixture of all kinds of languages.
You might hear someone say something like "avez vous finito mitten
der mufta," which is French, Italian, German and Arabic all in one
sentence, meaning, "Are you finished with the wrench?"

Every two weeks Jerry and I took the company plane back to Tripoli for one week of field leave and two other geologists replaced us. During our field leave we submitted our logs and reports and then had the week to ourselves. Jerry went home to his family and I joined the other Oasis Oil bachelors for our usual activities, such as snorkel and scuba diving, golf, sidewalk cafes, and dining at Italian restaurants.

I joined the Tripoli Yacht Club in picturesque Tripoli harbor, which was operated by the British Army. The club had a fleet of small 14-foot sailing dinghies, called "GP-14s," which had a jib sail up front and a main sail behind. They served a good curry lunch on Sundays too. I had to take lessons for several weeks before they would license me to take a boat out on my own. With one person aboard and a stiff breeze these little plywood boats would really lean with the wind and move fast. Sailing alone was a great passion of mine and I really enjoyed it. Then I made the mistake of deciding to buy a locally made sailboat that was for sale at the club. It was about the same size as a GP-14 but heavier and with a very tall mast. My boat was anchored at the Yacht Club, which was located in one corner of the harbor. The big problem was that during heavy storms from the north the waves in the harbor concentrated on the Yacht Club and really tossed the boats at anchor around. For some reason my boat leaked and was easily swamped, and I sometimes came to the club to find only the mast sticking out of the water. I had to pay the boat-keepers to raise it and pump it out.

After a week's leave in Tripoli, it was back to the drilling rig in the desert for another two weeks. As the well drilled deeper and deeper, we encountered a few minor shows of natural gas, which was encouraging.

In May 1958, as I was sitting in my office in Tripoli, I received a letter from Jim Knudstad, my old high school friend and archaeologist. Jim wrote that he had been assigned by his employer, the Oriental Institute of the University of Chicago, to an archaeological dig at the village of Tolmeita in eastern Libya. This village is located on the Mediterranean coast of the Libyan Province of Cyrenaica, about 75 miles northeast of Benghazi. Jim invited me to fly over and join his dig for as long as I wanted to stay. Excited at the thought of archaeological adventures in Cyrenaica, I arranged to take some time off to visit Jim on my next field leave from my geological duties on the Bahi well.

On May 31, I flew over to Benghazi on a Misrair (Egyptian Airline) flight and took a taxi to the Oasis Oil Company office. I had lunch with our resident manager and then he very kindly lent me his personal car for my trip to Tolmeita. I checked into the old and rather impressive Berenice (the old Roman name for Benghazi) Hotel on the harbor front, had a nice dinner and a drink in the rather seedy nightclub in the basement, which everyone referred to as "the snake pit." The next morning, I drove northeast out of Benghazi along the coast road and then up through the Gebel Akhdar (Green Mountains) by way of the Tocra Pass. I recalled seeing World War II movies of German General Field Marshal Rommel's Panzer tanks and trucks crawling up this same Tocra Pass. Then, I arrived up on the Barce Plain, famous for its nice climate and wines. Some miles later, I descended steeply down from the high plateau to the sea and then arrived at the little village of Tolmeita.

Using Jim's directions, I found the ancient ruins and the small concrete house being used for the expedition field quarters. Inside, I found Jim Knudstad and he introduced me to the small expedition staff. In charge was a distinguished archaeologist named Dr. Carl H. Kraeling, and as his second in command was the well-known Dr. Charles Nims, who was acting as the photographer on the dig; their wives were also members of the staff. Also on the staff were Mick and Pauline Wright. I was shown the small cubicle that was to be my room, stowed my gear away, and had a simple dinner with the staff.

Jim and I were up at 6:00 a.m. the next morning, and after breakfast he gave me a tour of their excavations and the surrounding Greco-Roman ruins. Jim explained that Tolmeita is the site of the ancient Greek and Roman (Greco-Roman) seaport of Ptolemais. The city was founded around 300 BC by Greeks under the Egyptian Ptolemy Kings, and thrived on exports of grain, olive oil, wine and silphium (a rare and valuable medicinal and cooking herb). It was taken over by the Romans in 96 BC and for a time the region was administered by Marc Antony for Queen Cleopatra of Egypt. Jim showed me "the glory hole" where they were currently excavating a Roman bathhouse, including the remains of a furnace that had once heated air to be circulated under the floor to form a sort of sauna bath.

Then, Jim showed me the most prominent feature of the ancient city, which was the ruins of a large city gate that was built about 350 BC of large limestone blocks. This elaborate gate, which may have been seven stories high at one time, had been gradually covered with earth over thousands of years. The gate had now been excavated and this revealed graffiti carved in the stone that changed languages upward as the gate had been slowly buried. The graffiti started with Greek words at the very bottom and then progressed upward into Roman Latin and then on up to Arabic, then Italian, and finally to English at the top from World War II British soldiers.

We went back to the excavation site and I took slides and movies of the expedition's work, including some very fine statues that had been unearthed. Jim's job, as a graduate architect, was to make architectural drawings of the excavations and he showed me some of his work on a Roman villa. Then, we toured through the small Tolmeita museum that had some nice specimens of Greek and Roman pottery and statuary. That afternoon, Jim and I did some snorkeling around the old Tolmeita harbor and saw some good-sized fish. We also searched in vain for a large statue that was said to be underwater near the ancient lighthouse and to be shaped like bird's wings.

The next day, Jim and I got into my car and we headed east to visit the spectacular ruins of the Greco-Roman City of Cyrene. We drove though Wadi Kuf, Beda, and Shahat to the ruins of Cyrene. Greek settlers founded this ancient city in the seventh century BC. In 331 BC it was conquered by Alexander the Great of Macedonia. Later, it was part of the Egyptian empire and finally it became a province of the Roman Empire in about 285 BC.

Cyrene was built around a spring in an eroded river valley at an elevation of about 1,800 feet above sea level and on the north side of a mountain range that sheltered it from the desert winds. It prospered due to good climate and soil, exports of silphium, a good supply of spring water, and precious metals; and, it is even mentioned in The Bible (Simon of Cyrene). On the terrace of the hotel overlooking the impressive ruins, we met with Dr. Richard G. Goodchild, the British Director of Antiquities for the Province of Cyrenaica, and then we toured the ruins. The two of us were especially impressed by the

Temple of Zeus, Sanctuary of Apollo, and the many tombs carved into the mountainside.

Back in our car, we drove down a winding mountain road to the tiny village of Susa on the Mediterranean coast. Susa is located at the site of the ancient Greco-Roman port of Apollonia, which had loaded goods to and from Cyrene thousands of years ago. We checked into the small but comfortable Dolphin Hotel, which had pretty gardens and a patio with a nice sea breeze. Jim and I then inspected the ruins of Apollonia. Especially notable were the remnants of the old city gate to the harbor and the ruins of two Byzantine (late-Roman) Christian churches, which were under excavation by Libyan government archaeologists. It was obvious to us from walls extending into the water that most of the ancient harbor was now under shallow water near the shoreline. We eagerly donned our snorkel gear to see what was underwater. Peering through our diving masks, we saw remnants of stone walls that had once sheltered the harbor, a scattering of building foundations, which had probably been warehouses, and numerous Roman columns. We decided to draw a sketch map of the old harbor and started methodically examining the old port facilities both under water and on the beach.

We could see submerged slips, where ancient ships had once been anchored and dry-docked, extending from the side of a small island down into the water. We also found the submerged harbor entrance channel filled with sand and bounded by remnants of stone walls. We spent the night in our hotel and had a pleasant dinner on the patio with a nice sea breeze. The next day we snorkeled around the rest of the small, present day harbor outside of the submerged walls of the ancient harbor. In addition to Roman columns, we saw: large crayfish (clawless lobsters); two very large manta rays about six feet across; many five-inch artillery shells from World War II; ancient pottery fragments; and two old ship's anchors with short lengths of anchor chain attached.

We finished our sketch map on the back of my Oasis Oil Company World War II minefield map of Libya. Then, we packed up and drove back up the winding mountain road to Cyrene, where we again visited with Dr. Goodchild and also toured the impressive sculpture museum. For a picnic lunch, we stopped at a tiny roadside shop and bought bread, two cans of corned beef (one was spoiled), onions, and a couple

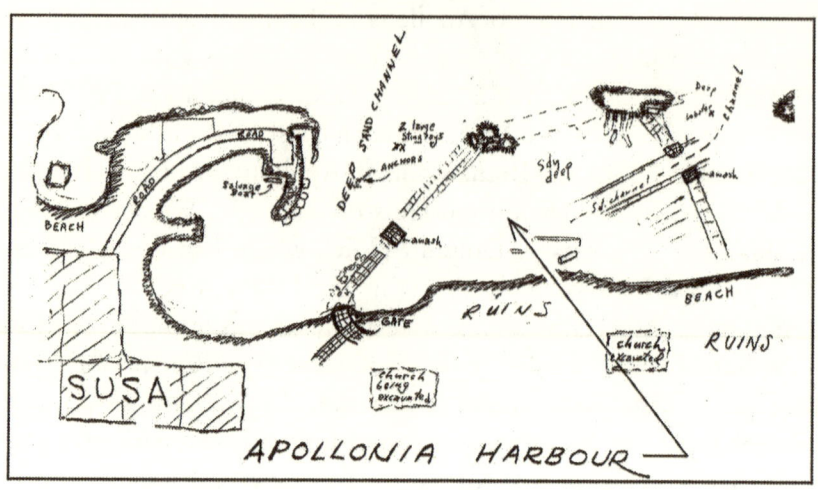

**OUR SKETCH MAP OF THE ANCIENT GREEK-
ROMAN PORT OF APOLLONIA.**
Using snorkel gear for several hours, Jim and I sketched out
this rough map of the ancient harbor of Apollonia. Parts of
the old harbor wall were awash and we could see the rest under
shallow water, including the channel into the harbor.

of bottles of soft drinks. After lunch, we continued our inspection of
the interesting ruins of Cyrene.

Having completed our tour, we drove back to Tolmeita, where I
had a final dinner with the expedition staff. After dinner, I drove
back to Benghazi and spent the night in the Oasis Oil Company
bachelor quarters. The next day, I boarded a British European Airway's
Elizabethan-type aircraft and flew back to Tripoli. It had been an
enjoyable and adventurous trip to Cyrenaica, and the ancient ruins of
Tolmeita, Cyrene, and Apollonia. And, it was amazing to contemplate
that two 1949 graduates of Clayton High School in St. Louis, Missouri,
had just been snorkeling around together over submerged ancient ruins
in Cyrenaica, Libya, North Africa. This had been just the kind of
adventure that the two of us had dreamed of as we had sat bored behind
our high-school desks some nine years earlier!

I flew back to Tripoli from Benghazi and then flew by company
plane back out to the Bahi wellsite in the desert. The well was
continuing to drill at a depth of about one-mile (about 5,300 feet)
under the ground.

Then, late one evening in mid-June 1958, Jerry came to our living trailer and asked me to come to the geological trailer. He said that the well had hit a fast drilling break and he had stopped the drilling to circulate up the rock samples. He said that there were strong traces of brown oil in the rock cuttings and oil floating on the mud pit. This was exciting news! As geologist-in-charge, I examined the rock samples and confirmed the oil shows. I then ordered the driller to go into the hole with a special tool called a "core barrel" to cut a 50-foot-long solid column of rock, called a "core," for detailed analysis. Finally, I called our Tripoli office on the short-wave radio and told them what was going on. They were quite excited and asked me to radio them again as soon as the core was cut and pulled out of the hole. Could it be that Oasis Oil had struck oil on its first well in Libya against all the odds?

Jerry and I went to bed after giving the driller orders to wake us when the core was reaching the surface. Some six hours later we were awakened and went to the rig floor. We had our rock hammers ready and the crew had laid out six-foot long, wooden core boxes on the rig floor that were numbered in sequence. The core barrel came out of the hole and the crew slowly started letting the three-and-one-half-inch diameter rock column out onto the rig floor. Jerry and I broke it into one-foot sections with our hammers and laid the sections in the open core boxes. After all 50 feet were laid out the crew started wiping off the core sections with rags. We immediately saw that we had porous sandstone and that gas was bubbling out of it; these were very good signs. Next we broke open sections and saw it was saturated with brown oil stain, which was also very good. Next we tasted the rock to see if it tasted of salt water, which would be bad, but there was no salty taste. We definitely had a very promising oil zone, and we proceeded to make a detailed description of the 50 feet of rock. We got on the radio and reported our exciting news to Tripoli and told them we were going to analyze the core using our laboratory equipment and recommended that we carry out of "drill stem test" to test the oil potential. Tripoli concurred with our recommendations.

We broke off pieces of the core and took them into our geological trailer. We got out our canning equipment and sealed pieces of the core into tin cans for complete analysis back at our Tripoli laboratory. Then, we used a rock drill to cut small plugs out of the core. We ran compressed air through these plugs to measure how permeable the sandstone was, and then used a mercury pump to force mercury into plugs to measure the total porosity of the rock. Finally, we put plugs into an oven to cook off all the oil and water inside to measure the oil saturation. All our tests indicated a potentially commercial oil zone and we radioed this information to Tripoli.

The next step was to run a "drill stem test" (DST) of the rock layer to determine the oil flow potential. I instructed the rig crew to rig up the testing equipment. I examined the drilling rate log to find the top of the oil zone and told the crew to place the "packer" at this point. The packer would go on the drill pipe above the oil zone and by putting weight on it would expand and seal the hole off above the oil zone so no water or rock cuttings from up the hole would come in during the test. A bottom hole pressure gauge was inserted at the bottom of the drill pipe and the pipe was then lowered down to the bottom of the hole. We waited for daylight so we could shut off all the electricity in the camp to avoid a fire hazard. The driller then eased off his brake and put weight on the DST tools at the bottom of the hole. The packer expanded and squeezed off the upper hole and the drill pipe was rotated to open the valve below the packer to allow whatever was in the oil zone to flow into the drill pipe. On the rig floor, the driller attached a small hose to the drill pipe and put it into a bucket of water. We all stood around the bucket and watched as it started to blow air bubbles, which indicated that something was flowing into the drill pipe. Every few minutes I smelled the hose to see if the air had turned to natural gas, which would be a very good sign.

Finally, my nose told me that we had gas to the surface and I recorded the time. We then hoped we might have oil flowing to the surface, which would mean a commercial oil discovery. But, after an hour or so the gas bubbles in the bucket slowed down and indicated that whatever was coming into the drill pipe had almost stopped. We decided to stop the test and see what we had recovered in the pipe.

The driller and roughnecks started coming out of the hole and we could see gas fumes as they unscrewed each section of drill pipe. Then, as we got to the last several hundred feet of pipe, good brown oil poured out all over the rig floor every time they unscrewed a joint. We counted the joints with oil and calculated that we had recovered a quantity of sweet (not sulfurous) crude oil in the bottom of the drill pipe. Using a gauge, we found the oil was a good grade (not too heavy or too light).

Finally we got to the bottom of the pipe and the crew removed the bottom hole pressure gauge. We took the chart into the geological trailer and consulted our drill stem test manual. The gauge only left a scratched pressure line on a blackened copper plate with no calibration marks of any kind. We measured various line segments with a micrometer and using a manual converted them to pressure in pounds per square inch. The pressures were relatively low, and this, together with the small amount of oil recovered, indicated that the oil zone we had just tested had a relatively low potential. The A1-32 well was abandoned in July 1958.

Although not commercial, the Bahi was extremely encouraging because we had found oil on our first well and this indicated a very large oil potential under our vast Libyan acreage.

Following our drill stem test on the A1-32 Bahi well, I went to town and spent the next two months working in the Tripoli office

CRUDE OIL IN THE TANK ON THE A1-32 DISCOVERY WELL.
Members of the crew celebrate oil in the tank after testing
our first oil discovery in Libya. On top of the tank are our
British doctor, a Libyan, and my colleague, Jerry Hayes.

on various reports and maps. Our department also went to work planning our second well location in Concession 32. As September 1958 approached, I made final preparations for my carefully planned vacation trip to Uganda, Kenya, Tanganyika, and Zanzibar in East Africa.

12 ✴

East Africa and Zanzibar Trip Diary 1958

On September 14, 1958, I departed Idris Airport in Tripoli, Libya, via a Misrair (better known as "Miserable-Air") Egyptian Airline Viscount aircraft at 4 p.m. Arrived at Benghazi, Libya (Benina Airport) about 2 hours later. After dinner courtesy of Misrair, I boarded a BOAC Argonaut aircraft with four propellers and took off at 8 p.m. with the runway lit by kerosene lamps. Our plane cruised along at 11,000 feet southeast passing over Gebel Aweinat (Aweinat Mountains) located in the southwest corner of Egypt. From Gebel Aweinat we flew directly to Khartoum, Sudan, where we landed at 2:45 a.m. Khartoum is located at the junction of the White and Blue Nile Rivers. Boys wearing long white robes with red sashes and turbans served refreshments in the airport.

Departed Khartoum just before daylight after a one-hour delay while they changed a magneto in the old Argonaut. In the east first Venus and then the sun rose over the Nile River. The plane followed the White Nile River on south over Malakal and Juba, both in Sudan. Large irrigation systems for the cotton plantations were seen in Sudan by first light.

Day 2. Crossing over the border into Uganda, the country was very green and becoming hilly. Uganda was still a British colony. We flew over the Victoria Nile next and almost in sight of Murchison Falls. Passing over Masindi we arrived at the town of Entebbe on Lake Victoria.

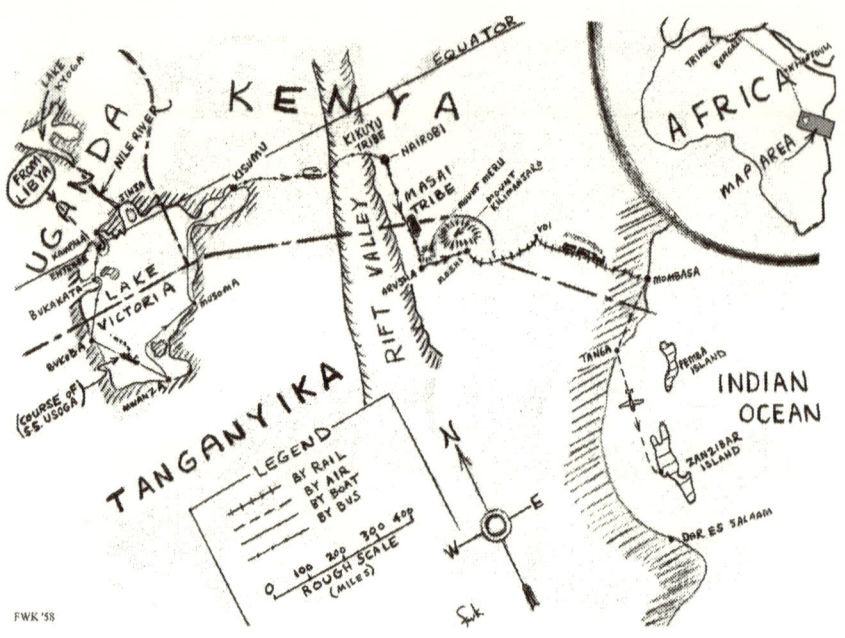

INDEX MAP TO MY 1958 EAST AFRICAN ADVENTURE.
This hand-drawn map shows my route: around Lake Victoria
by steamer, by mini-bus to Nairobi, local bus to Arusha, by
train to Mombasa, and then by plane to Zanzibar.

As the plane set down on the runway the first things I noticed were giant red anthills about six feet high along the airstrip. After clearing customs, I got on a BOAC Volkswagen Bus, and, along with the crew from the plane (a nice bunch of chaps), we drove north about 20 miles to Kampala, the capital of Uganda. I checked into the Imperial Hotel, the best hotel in town. Kampala is very green and has a pleasant climate, being about 4,000 feet above sea level. The city is built on a series of hills. The native Africans here are of the Buganda Tribe and speak Luganda. There is a large Indian population here also (in all East Africa I was to find out) and they own most of the shops. The Africans wear colorful clothing and appear happy and friendly. My room in the Imperial was very nice and meals very good.

Day 3. This morning I was introduced to an untimely British custom of East Africa. My alarm was set for 8:30 a.m. but at 6:30 a.m. my African roomboy, Mwenge, knocked on the door and walked in with a tray of hot tea. All this was too early for me. After breakfast, while waiting for a car I had hired to see the Kampala sights, I met a fellow who was a courier with the American State Department. He joined me and we toured the Kampala area. We visited: The Old Fort, Palace of the King of Buganda, Tomb of the Buganda Kings, a couple of churches, Mosque of Jamath Khana (followers of H.H. Aga Khan), Bulange (Govt. buildings), and Makerere College. Returning to the hotel I was amused to see African boys with a rag under each foot skating around the halls to polish the floors.

Day 4. Went by car and driver east about 50 miles to Jinja, Uganda, passing through Mukono, Buikwe, and other small villages on the way. At Jinja I viewed the impressive Owens Falls Dam, which is on Lake Victoria; the water flowing through the spillway is the beginning of the White Nile River. Saw a couple of hippopotamuses swimming near the dam.

OWENS FALLS DAM ON LAKE VICTORIA, UGANDA.
The water from this dam forms the beginning of the White Nile River
that flows north to the Mediterranean Sea almost 3,000 miles away.

After leaving the dam I noticed the buildings of Nile Breweries Ltd. set back from the road, so we drove in and I inquired about looking around the place. They were delighted to have a visitor and gave me the grand tour. After I had been through the brewery, they took me back into the refrigerated sample room where the assistant brewmaster almost got me drunk trying his many special mixtures of beers (pasteurized, unpasteurized, with preservative, without preservative, etc.).

Lunch in Ripon Falls Hotel overlooking the Napoleon Gulf of Lake Victoria. Drove back to Kampala through plantations of sugar cane, tea, bananas, coffee and paw-paw. Passed through Mibira Forest with jungle over 100 feet high along each side of highway. Vegetation here in Uganda is said to be similar to the Belgian Congo of West Africa.

Day 5. Drove to Port Bell on Lake Victoria and boarded the old, steamship, SS *Usoga*, which was to be my home for 4 days; I had reserved a berth from Libya using an old East African guidebook. The British hauled this boat up to the lake in pieces on the railroad just after the year 1900. After loading a large cargo of Bell Beer we steamed away to arrive at Entebbe Pier at 2 p.m. Again we took on cargo and a large crowd of Africans came to watch. The SS *Usoga* steams counter-clockwise around Lake Victoria, which is the second largest fresh-water lake in the world, being 150 miles wide and 250 miles long.

Departing Port Bell, we steamed slowly along (maximum speed on this 50-year-old boat is 7 mph) south towards the Equator, which runs right across the northern end of the lake. I humorously asked the British captain if we were going to have a ceremony as we crossed the Equator and he said sternly, "We cross the Equator twice a week and would never get any work done if we celebrated the crossings." We crossed the Equator without ceremony and then steamed south down through Salisbury Channel and Bugoma Channel past the Sese Islands. At 9 p.m. arrived at Bukakata. Here on the wharf a mountain of 290 tons of coffee beans was waiting. From 9 p.m. until 4 a.m. the ship loaded coffee with its ancient steam-winch (and my cabin was next to the hold; amazing what a person can sleep through).

Day 6. Steamed into Kagegi Gulf and then arrived at port of Bukoba, Tanganyika, at 3 p.m. Another stack of coffee was waiting on the quayside. Went ashore and walked into the village a few miles away. Saw H.M. (Her Majesty's) Prison, markets, etc. Stopped at the bus depot.

Some pretty wild and interesting people stepped off those buses. While I was at the depot, a wedding procession came by composed of young girls singing and clapping. Bukoba is located at the foot of a granite escarpment several hundred feet high that is probably formed by a fault line. Saw a group of otters swimming in the lake.

Meanwhile back at the boat, cargo loaded (we now have 500 tons of coffee aboard valued at $240,000). Took aboard a bunch of H.M. prisoners all chained together; they will sit on deck forward, which is where all the Africans ride.

THE S.S. *USOGA* DOCKED AT BUKOBA ON LAKE VICTORIA.
In about 1900, the British hauled this ship up to the lake
in pieces by railroad from Mombasa. At the time I took
this picture it was loading coffee and prisoners.

Another Westerner came aboard and moved into the other bunk in my cabin. He is a botanist and his name is Barrie Juniper (what a great name for a botanist). He is a student at Oxford and working on his doctor's degree in botany. He has just finished a 3-month expedition in western Tanganyika near Lake Tanganyika and is headed back to England. A very interesting fellow, he has a long black goatee.

Day 7. This morning the captain invited me up on the bridge, as I had previously requested. Main point of interest was the radar unit but also interesting maps and charts. At 11 a.m. docked at the port of Mwanza, Tanganyika. A large crowd was at the quay to meet the boat, as usual. Juniper and I, each with binoculars on neck straps, went ashore and spent the afternoon walking around in the bush. We spotted:

a group of monkeys, many lizards (some with red head, orange body and blue tail), fish eagles, cormorants, and many flowers, which Juniper identified (and some he couldn't). Near the pier here at Mwanza is a prominent granite outcrop known as Bismark Rock; the Germans had a beer-garden by these rocks back when they owned Tanganyika before World War I. Mwanza is the center of a large, rich cotton area; and, the famous Williamson Diamond Mines are located about 100 miles from here.

Day 8. Steamed from Mwanza and arrived at the port of Musoma, Tanganyika. Walked into the village and saw many interesting markets and dukas (shops). Departed Musoma 11:30 a.m. and just after clearing the port we passed the SS *Rusinga*, sistership to the *Usoga*, which goes around the lake in the opposite direction (clockwise).

About the other passengers on the boat: no other tourists, most of the Europeans are missionaries (doctors, nurses, and chaplains), but a few are plantation owners. One missionary told me a humorous story about a new missionary that traveled way up the Congo River to visit his new flock and was amazed to be served ham at every stop; what he didn't know was that the ham was traveling with him on his boat! Quite a few Indians on board too. Our meals in the captain's cabin are nicely prepared. The whole trip very inexpensive too. I ate at the captain's table (there were only 3 tables).

Day 9. Arrived at the port of Kisumu, Kenya at 7 a.m. and after breakfast I disembarked the ship to start my road trip to Nairobi. Got on a Volkswagen Microbus with two missionaries and two Indians and started down a dirt road to Nairobi. We drove down through very pretty, green mountainous country into the Rift Valley. The Rift Valley is a long, narrow valley across East Africa with steep sides formed by major faults. While in the valley we ran into a heavy thundershower. Our African driver, a poor driver under perfect conditions, kept right on at full speed in the rain and skidded dangerously on the wet pavement (gasp!). Then, ten minutes later he ran off the right-hand side of the road (they're supposed to be on left side in this country).

Drove through the Kikuyu Tribal Reserve. Members of this tribe made up most of murderous rebels during the Mau-Mau Rebellion against the British during the early 1950s; now all are in camps of two kinds. One kind, in which most Kikuyu live, has fencing and guard towers to keep Mau-Mau terrorists from entering and killing their own people; the other camps are where Mau-Mau terrorists are confined. Every Kikuyu woman I saw had a large load on her back carried by a forehead strap. Arrived Nairobi, Kenya at 3:30 p.m., safely but shaken after an exciting bus ride. Got a room at the YMCA at a very reasonable rate ($2.50 per day including all meals).

Day 10. Spent day looking around Nairobi and doing a little shopping, including a six-foot-long Masai spear and a colorful Masai woman's necklace. In the evening while sitting at the bar in the New Stanley Hotel met a young English fellow who is a minerals prospector for the Kenya Government. He invited me to his government-housing unit for dinner and gave me several souvenirs of Kenya. A very nice chap, as were most of the British colonials I met in East Africa; he made me his guest and was happy to have a new visitor.

Day 11. Today took a tour around Nairobi National Park, which covers a huge area and is not fenced on one side to allow the animals to migrate in and out. Saw herds of zebra, wildebeest, impala, and giraffe, and many specimens of ostrich, jackal, baboon, warthog and gazelle. The driver of the car, which I hired for a few hours, spotted a pride of 5 young lions. He drove up to within 20 feet of them and we sat and watched them for an hour or so.

Day 12. Boarded a 2nd Class bus at 8:30 a.m. The bus was divided into two sections: the forward half was for Europeans and Asians (as the Indians are called), while the rear 3rd Class was mostly for Africans. I was one of three Westerners, with the other two being Greek and Austrian; rest of bus filled with Asians wearing turbans, beards and other traditional Indian garb and the rear filled with Africans mostly of the Masai tribe. One Indian was carrying a four-foot-long sword, another a homemade battle-ax. Our battered, old diesel bus crawled out of Nairobi at 9 a.m. Fifteen miles out we developed a fuel leak and stopped for repairs. Continuing, we passed many animals on the roadside as we headed south into Tanganyika. We entered the Masai tribal area and saw many Masai along side of the road tending their cattle. These people are the most interesting that I have seen. They live entirely off their cattle, that is on meat, milk and blood. The women shave their heads and wear colorful necklaces and immense earrings some 8 to 10 inches in diameter; some wear anklets and bracelets of polished steel up to 6 inches wide. The men are tall and thin, set their hair with red clay, carry spears over 6 feet long (I brought one back to Libya with me) and a short sword, extend their ear lobes down with weights and wear an orange robe tied over one shoulder (as do the women). This tribe is very wealthy and they measure their wealth only in cattle; their herds are very large.

TWO MASAI WOMEN AND A MASAI MAN IN SOUTHERN KENYA.
I took this picture out of the window of the public
bus I was riding south to Tanganyika.

Crossing the border into Tanganyika we soon arrived at our destination for the day, Arusha. Arusha is located at base of Mt. Meru, which has an elevation of about 15,000 feet at its summit. Here I ran into a friend of one of the Oasis employees in Tripoli; also I met a fellow who is a veterinarian with the British government; learned much about the area from this "vet" and had him over to my hotel for dinner. Mighty friendly bunch of people here in East Africa; they are always glad to see a new face come into these little towns. Spent the night in the least expensive hotel available.

Day 13. Re-boarded the same bus as the previous day for 2-1/2 hour ride to Moshi, Tanganyika. Bus is very full of interesting Africans and their bananas and paw-paw. Arrived at Moshi at 10 a.m. to see a sight I had looked forward to seeing for a long time. Behind Moshi rose the mighty Mount Kilimanjaro with her snow-capped peak shining at 19,340 feet above sea level, the highest point on the African continent. A beautiful sight and well worth the trip from Nairobi. I had hoped to climb to the top of Kilimanjaro, but time would not permit (climb takes a minimum of 5 days). Spent the night at the Livingston Hotel with a beautiful view of Kilimanjaro (I could even see the snow cap in the bright moonlight at midnight).

Day 14. Spent morning poking around in the local markets and shops. At 5 p.m. boarded an East African Railway train consisting of two small locomotives, three old coaches and a few tank cars. A fascinating group of Africans was assembled at the station. With a jolt, I was on my way to the port of Mombasa on the Indian Ocean. A Scotsman from South Africa shared a compartment with me and we were the only Westerners aboard. The train stopped at every little siding and village (such places as Rau River Halt, Tavota and Ziwani) and then we stopped at a little rest hut at Maktau Halt for a dinner of fish and chips. At Voi our coaches were hooked onto the main first-class train from Nairobi, and then the whole works continued on to Mombasa.

Day 15. The train crossed a causeway onto Mombasa Island in the Indian Ocean. Checked into the Tudor House Hotel overlooking Port Tudor on the north end of the Island.

In the afternoon took a city bus to Fort Jesus, built by the Portuguese in 1593. The fort is located on Old Mombasa Harbor which is used now only by the Arab dhows which come from the Persian Gulf, India, and the Gulf of Aden on the monsoons which blow from January to April. While it was not dhow season, I walked through the narrow streets of the Arab Quarter and found one colorful dhow tied near shore. The narrow maze of streets in the Arab Quarter was very interesting with old customhouses, fish markets, etc.

Day 16. Spent the morning walking around Mombasa. On the way to the modern harbor, I walked through a double arch erected over the main street (Kilindini Road) to greet Princess Margaret on one of her visits. The arch consists of huge steel crossed elephant tusks painted ivory color. After clearing with the port authorities, I went into the modern Mombasa harbor known as Kilindini Harbor. This is the most important port of East Africa and has modern wharves and protected anchorage.

Spent the morning watching the shipping. With the shipping page out of the local newspaper and my binoculars, I could identify each ship as well as her cargo and destination.

Walked back into town, toured around and did a little shopping. A strange mixture of people here consisting of English, Indians, Africans, and Arabs, as well as the mixed Arab-Africans known as Swahilis. Clothing colorful. Many wear a *kikoi*, a bright cloth wrapped around the waist, Swahilis wear white robes with embroidered vest and cap. Arabs wear a cloth around the waist and a cloth wrapped around the head in a crude turban, and some have embroidered robes. Women wear all kinds of colorful cloth draped around. The Moslem women wear a black garment, which covers their face, and the Indian women wear colorful silken *saris*.

Day 17. Boarded an East African Airlines DC-3 at 10:00 a.m. to fly to Zanzibar. Flew down the coast southward again crossing into Tanganyika. Landed at Tanga, Tanganyika for 20 minutes. I was amused that all automobiles in Tanganyika have an identification plate that says, "EAT," which stands for "East Africa Tanganyika." Taking off from Tanga, our plane flew out over the Indian Ocean passing many small islands.

At noon our plane circled over the picturesque island of Zanzibar and landed at the airfield. Checked into one of the two small hotels on the island. Zanzibar is an island about 53 miles long and is the farthest south latitude (about 6 degrees south) I have reached so far in my travels. After wandering blindly through the maze of narrow streets that make Zanzibar town (the main street is about 15 feet wide), I rented a bicycle and started cycling out north of town. The palace of the Sultan of Zanzibar was the first place of interest I visited. Above the palace flies the solid red flag of H.H. (His Highness) the Sultan. Next, I passed Funguni Spit, which is a sand spit extending offshore covered by crude dwellings. Repairs to dhows are carried out on this spit. As it was not the dhow season, all I saw on this spit was a row of old rotting hulks of abandoned dhows.

Next, I cycled past a house known as "Livingston's House," where the famous Scottish explorer and missionary, Dr. David Livingston, lived while fitting out for his last expedition to the mainland in 1866. Before this, Livingston had explored all over central and east Africa. After leaving

Zanzibar in 1866, Livingston disappeared into the "Dark Continent" until he was discovered years later by journalist, Henry Stanley, who greeted Livingston with his famous words, "Dr. Livingston, I presume?"

Moving on north, passed H.H. (His Highness) Aga Khan Club and came to the ruins of Marahubi Palace about 3 miles out of town. Nothing much to see except mango trees, stone aqueducts and some blue waterlilies. Continuing on, about four miles out were the ruins of another palace built before 1800 by some Arab potentate, as was the Marahubi. The place is now a government petrol *godown* (local lingo for a warehouse). About 4-1/2 miles out, I had had enough pedaling (considering the return trip). Returning to Zanzibar town, I visited the Beit al-Amani or Peace Museum.

Day 18. In the morning walked to the harbor and watched fishermen hauling nets and sailing about in their outrigger dugout canoes. Rented another bicycle (cost about 25 cents for three or four hours) and cycled southeast from town through plantations of coconuts and cloves. Watched natives with a rope tied between their feet climbing the coco-palms and cutting the coconuts off. They would yell each time that a cluster of nuts would fall to the ground.

Many clove trees along the road. Zanzibar is the clove center of the world. Cloves are dried unopened buds from the clove tree. About 5 miles out turned around and headed back to town. Cycling back I spotted some men chopping a dugout canoe out of a felled tree. Went over and watched for a while. Spent some time looking around Zanzibar town and did a little shopping. Examined several of the famous "Zanzibar doors." These doors are large, heavy and ornately carved. They are vertically halved doors and each half is decorated with pointed brass bosses; it is thought these points originated in India where they were used to fortify doors against war elephants. The intricately carved uprights and lintels are the main features.

Day 19. At 2:30 p.m. boarded an East African DC-3 and flew back to Mombasa by way of Tanga. After leaving Zanzibar my outward journey was finished and I retraced by plane back to Nairobi by way of Moshi (beautiful view of Kilimanjaro sticking up through a cloud layer) and Arusha (rough part of the flight; the airsickness bags were doing a good business).

Day 20. Boarded a BOAC Argonaut at Nairobi. Took off two hours late while they repaired the feathering device in one engine (Argonauts are to be taken out of service soon fortunately). Landed in Khartoum again at 10 p.m. and the temperature was 92 degrees.

Day 21. After a stopover in Benghazi, returned to Idris Airport, Tripoli, Libya, safe and sound on October 4, 1958, after an approximately 7,500-mile trip. Looking back, I had to say that East Africa was everything I had expected, and much more. Something new and adventurous every minute!

13 ✴

Black Gold in the Sand

Upon my return from East Africa, I was assigned to head up a shallow, core hole, drilling project in our Libyan Concession 25. This project, called the Uadi Matab Corehole Project, involved drilling a series of shallow holes on a good drilling prospect to try to confirm that the favorable surface geology looked just as good below the zone of surface weathering and slumping. These holes were called "coreholes" or "stratigraphic holes," because they were intended only to examine the underground rock layers and not to look for oil.

In November 1958, we convoyed a small truck-mounted drilling rig, a house trailer and a supply truck to the drill site and I surveyed in the first well location. Our camp was located near the oasis of Bu Gnem. I drove over to this picturesque oasis of palm trees surrounded by high sand dunes and was impressed by the ruins of an old French Foreign Legion fort, shown as "Fort Demante" on my maps.

George Frietag, who had been a German SS officer during World War II, was assigned to our camp as camp boss and to take care of land mine clearance. Every morning he jogged around our camp several times. I asked him why. He said he wanted to be in shape when World War III started! One evening I was talking with Frietag about World

War II, which had ended only 15 years earlier. He said he had fought on the Russian Front and then was transferred to the Western Front after the Allied armies landed in Normandy, France, on June 6, 1944. I said to Frietag that Germany had started World War I and World War II and would start another war if it could. Frietag said this was not true because the Americans had troops in Europe, the French and Dutch had rearmed and the British military was strong. I replied, "You see, George, you're not saying that Germany would not want to start another war, only that it cannot start one."

One day I found Frietag out in the desert near the camp tearing apart a German S-mine from World War II. The S-mine was an anti-personnel mine, which when stepped on shot six feet in the air and then exploded shooting steel ball bearings in every direction; it was called a "Bouncing Betty" by the American forces. Frietag had opened this mine and was counting the ball bearings. Then he said to me, "This is why Germany lost the war. Defective munitions. There should be 350 ball bearings in this mine and there are only 325."

Another evening Frietag told me about the time he was eating in one of our desert camps a couple of years earlier and was introduced to Earl Bauer, Ohio Oil Company's aviation manager from Findlay. Bauer started telling the story about how he had been a fighter pilot in World War II and was shot down over Germany. He said he parachuted out of his plane and landed in a farmer's field. All the farmers close by grabbed their pitchforks and other tools and rushed over to him and were going to kill him on the spot. Then, he said, a German officer drove up, drew his pistol, and took him prisoner and saved his life. Frietag then said to me that he leaned across the table and told Bauer that in fact he was the German officer that had saved his life in that farmer's field!

Some days later, I asked Frietag to teach me how to do the Nazi marching step, known as the "goose step." He showed me that there was more to it than it appears. The back foot rises on its toes as the front foot projects and kicks sharply forward like a ballet step. George was not at all perturbed about all my questions about his German army days in World War II, and, in fact, seemed to enjoy talking about his "good old days."

We also had an American driller in our camp. He was of Danish extraction and upheld the reputation of many drillers by being rather crude and vulgar, especially with his Libyan workers. When he moved their small drilling rig to a new location and his Libyan workers started to pray toward Mecca in the east, he would tell them they were praying in the wrong direction, just to upset them. He only knew one phrase in Arabic, which was "stick it up your rear," and no matter what his Libyans complained about they always got this same answer.

Also in our camp, we had a Libyan manager who was in charge of the Libyan workers. As our camp was quite small, this Libyan was allowed to eat with Frietag, our driller, and me in our small combination kitchen-dining room trailer. One evening, our cook served us all a plate of some kind on non-descriptive meat covered with gravy. As we started to eat, our driller said, "this is sure good pork we are eating." As Moslems are forbidden to eat pork, our Libyan choked and excitedly said, "No, this is not pork, this is beef." The driller then said, "No, I was raised on a farm and I know pork when I see it." The Libyan again insisted it was beef. Finally, the driller said to the Libyan, "This is pork all right, and if I am not mistaken the piece you have came from right around the pig's rear end." As any reference to "rear ends" was also disturbing to Libyans; this really disgusted and sickened the Libyan and he pushed his plate away and said nothing more. Freitag and I just sat there and witnessed this whole bizarre scene without comment. [I have always felt such disrespectful and insulting behavior towards the Libyans helped lead to Libya's Colonel Gaddafi taking over the country in 1969 and kicking all the foreigners out.]

The Uadi Matab Project ended up in the drilling of five shallow wells ranging in depth from about 200 to 800 feet deep. I logged the rock samples using my microscope as each well was drilled. Then, after each well was finished drilling, I unloaded a Witco portable well logging unit from a Landrover and ran a gamma ray electric log to help correlate the wells along with my sample logs. I was able to run a complete electric log with gamma ray and resistivity curves in only one hole because the others encountered large cavities that drained away all the drilling mud, which was needed to log resistivity.

LOGGING A STRATIGRAPHIC HOLE.
I logged each of the shallow holes on the Uadi Matab
Project with a portable Widco electric logging unit.

After drawing a series of maps and cross-sections, I finally concluded that the shallow rock layers under the surface did indeed confirm the good drilling prospect that our geologists had mapped on the surface. This structure was scheduled for future drilling.

In late December 1958, I was working in our Tripoli office assessing Libya's overall oil potential. At the time we had no maps of the underground rock layers in Libya to work with because so few wells had been drilled. I decided to compile a structural map of a formation called the "Cambro-Ordovician Quartzite" that underlay the entire area of oil potential presently being explored by several oil companies. My map was finished on December 12, 1958, and was the first subsurface map created for Oasis. Our senior management took quite a keen interest in my interpretation and it was helpful in locating future drilling locations.

At about this time, I spotted a newspaper ad with a used, white, MG Model A, British sports car for sale. So, I sold my nearly new Volkswagen and traded up to a nifty sports car. It was great to be back driving MG cars again, and the weather in Tripoli was ideal for "top down" driving.

As Christmas 1958 approached, the question came up in Tripoli of which geologists would work on the drilling rigs during Christmas. Much to my chagrin it was decided that the married geologists would spend Christmas in town with their families and the bachelors would man the drilling rigs. Although I was due for field leave in town, I would now have to stay over for Christmas in the desert. This was the saddest Christmas I ever spent. We few Americans sat around in the mess hall feeling lonely and sorry for ourselves. The company had sent out a Christmas tree but nothing to put on it. We sat and looked at the bare tree and it made us feel even worse. Then the British doctor, who also had to stay at the rig, came up with an idea. He suggested that he could decorate the tree with cotton balls from his clinic, the electrician could wire together some headlight bulbs to string around the tree, and we could put candy bars under the tree. And, that is what we did. It was a rather pathetic effort reminiscent of the stories from the prisoner of war camps during World War II!

Meanwhile our second well in Libya had started drilling. It was called the "B1-32 Dahra," meaning it was the second well in our Concession 32 and located at a place on the maps called Dahra. I flew out to the well location in the desert and took up regular wellsite duties of 12-hour shifts for two weeks and then a week in Tripoli writing reports and time off. One day our C-47 company plane arrived with supplies. The pilot told us that on the way to the camp the Libyan rig workers on board decided to stop eating the Swiss cheese triangles we had been supplying them because they had decided that the picture on the round boxes was a pig and pork is forbidden for Moslems. The picture was actually a mountain goat but none of the Libyans had ever seen a pig and someone spread the word that the picture was a pig. It took some vigorous explaining to convince the Libyans to start eating the cheese again.

In our geological trailer, we started to record some large indications of natural gas on our detector. The rock cuttings, or "samples," did not look very encouraging, but it was decided to run a drill stem test to see how much gas was present. We went in with our DST tools and opened up the test. We had gas to the surface very soon but the pressure was weak. When we pulled the drill pipe out of the hole we recovered only salt water. While of no commercial interest, the presence of natural gas was a good sign that oil could be present at greater depths.

One day the generator failed and the lights went out while I was having dinner in the mess trailer. Several other Americans and I sat there in the dark waiting for the power to come back on. Then, a Lebanese mess-hall worker poured olive oil into a saucer, rolled up a paper napkin, laid it in the oil, and lit it with a match. We were amazed to see a very bright flame erupt from the dish, which lit up the whole trailer. Then, I remembered that was exactly how the clay lamps used by the Greeks and Romans had worked. An Arab knew this but Americans would never think of it!

The well drilled ahead and we drill-stem tested a couple more weak gas shows, but nothing of commercial interest resulted. Then, in March 1959, I was assigned to take charge of the D1-32 wildcat well about to be drilled in our Concession 32. Meanwhile, the B1-32 Dahra well had drilled on after I left and hit big oil to make it Oasis Oil Company's first commercial oil discovery. This well discovered the first giant oilfield in Libya.

After some leave in Tripoli, I took up my duties at the D1-32 wildcat (a wildcat is a well in totally unproven territory) well located in our Concession 32 about ten miles north of the abandoned A1-32 dry hole, on which we had previously tested a non-commercial oil zone. The D1-32 turned out to be a rather long and boring well.

Between electric-logging runs, our French logging engineer had absolutely nothing to do for weeks on end except some light maintenance of his equipment. To keep himself occupied, he had brought with him copies of all the Paris newspapers for one year. He was a mathematician and he used his math to analyze all the horse racing results in Paris for a year in order to try to develop a formula to predict which horse would win a horserace. He was a rather strange and quiet guy. He quit some months later to return to Paris and when he was leaving he told us that he had worked out a mathematical formula to predict horserace winners based on the jockey's weights. He asked us if we wanted to invest $1,000 each in his formula, but, needless to say, we turned his proposal down. I have always wondered if he became a wealthy minister of France or ended up in an insane asylum.

We drilled ahead without anything interesting. Then we received word that a field crew of British Petroleum geologists working deep in the desert in far eastern Libya had just discovered the wreckage of a B-24 Liberator bomber that had crashed during World War II. The plane was called the *Lady Be Good* and its discovery created a mystery that was followed by millions of Americans over the next couple of years. The U.S. Air Force investigated the wreck and discovered that the nine crew members had bailed out and their bodies were missing. As a result, the U.S. Army Mortuary System launched an extensive search of the area and the history of the plane was researched. It turned that the *Lady Be Good* had flown its maiden voyage on a bombing mission over Naples, Italy, in 1943 from an airbase near Benghazi, Libya. On its return, it overflew its base in the dark and mistakenly continued south into the desert until its fuel ran out. The nine crew members bailed out and the plane belly-landed in a sea of high sand dunes about 400 miles from the coast. Thinking they were near the coast, the crew walked north through the sand dunes until their water and food ran out. Remains of eight out of the nine bodies were eventually found in the sand. One of the crew kept a diary that sadly documented their demise.

Meanwhile, our well continued to drill without encouragement. Out of boredom, I started drawing pen-and-ink cartoons of life in an oil camp in the Sahara Desert. I also took up Solitaire and played an intricate version called "Grand Napoleon."

Finally, to kill time, my fellow geologist and I started working on a memory system I had read about. It started by making up a list of visual images for all the numbers up to 500 and assigning a letter for each of the numbers 0 through 9 that resembled the number. So, the number one was "t," two was "n," three was "m," etc. Thus, the number 23 was "nm" and vowels and silent letters did not count, so we would invent a word like "gnome" to be the number 23. After this was done, we worked towards memorizing a list of 500 random objects using our visual numbers. If the 23rd object was "bicycle," we would visualize a gnome riding a bicycle. We eventually memorized a list of 500 items and could recall any item at random. This was an example of the bizarre ways we passed time in a lonely oilwell camp in the middle of the desert.

ONE OF MY OIL DRILLING CAMP CARTOONS.
This cartoon was drawn on the D1-32 location. It depicts the occasions
where there were not enough empty seats on the return leg of the supply
flights to the camps and someone counting on going to town was left behind!

Then, as if things were not bad enough, one of the Libyans working
on the rig floor had his arm torn off by the machinery and died. An
American roughneck panicked and threw the arm into a drilling-mud
pit. The other Libyans started rioting because according to their religion
a dead body must be buried with all its parts. The America toolpusher
in charge of the drilling finally calmed them down by a mixture of
tough language and the promise that the arm would be recovered from
the mud pit. The arm was recovered and a plane flew down from
Tripoli to take the body back to the man's family for burial.

Our D1-32 well was finally stopped and proclaimed, "D & A," that
is "dry and "abandoned," as no oil or gas was found. In May 1959, it
was time for me to go on home leave to the States again. On my way to
the States, I decided to visit my old haunts in Munich, Germany, again
and then go on to the famous city of Berlin. I flew to Munich and
made the rounds of my favorite places. This included: venison and red
cabbage in the Rathauskeller (Ratzkeller) of City Hall; drinking wine
in the Grinsing wine district; drinking beer and eating dumplings in
the cellar of the Hofbrauhaus; watching the Klockenspiel clock strike

the hour with mechanical knights jousting; and drinking beer in the massive Matauser beer hall by the railroad station. I also went on a bus tour to the site of the Nazi concentration camp at Dachau, just outside Munich. This was very impressive, especially as I met a man on the bus who had been a prisoner there during World War II and was revisiting the place.

From Munich I flew to Berlin and checked into a bed-and-breakfast hotel that I found out about at the airport. I had quite a problem with the bed. It had a wedge under the head of the mattress, no upper sheet, and an eiderdown only six feet long that left my feet sticking out. I got to work and took the wedge out and then called the office and asked for two blankets. I wrapped the foot of the bed with the blankets and then all was well. I had heard about Berlin all my life and especially about the terrible bombing during World War II, the burning of the Reichstag, Hitler's bunker, and then the Russian Berlin Blockade and the Berlin Airlift. [However, this was some years before the Berlin Wall was constructed.] I decided to take a bus tour to see the sights.

After seeing the tourist places in West Berlin, our bus went through "Checkpoint Charlie" with its armed Soviet guards and then into the Russian Zone of East Berlin. What an eye-opener this was! Every building and factory was in total rubble just as it had been left after the WWII bombings; nothing had been touched. The only things the Russians had added were huge propaganda banners tacked onto the walls of the larger ruined buildings. We drove slowly by what had once been beautiful cathedrals and opera houses but were now completely destroyed.

Then we drove down the famous street leading to West Berlin called "Stalinallee" (Stalin's Alley). This wide boulevard was lined with large, ornate buildings that were always photographed in the background by the Russians when famous visitors were driven here. However, our bus driver then showed us that all the buildings were just false fronts with supports in the back! Needless to say East Berlin was dreary and depressing. It spoke volumes about the Soviet Empire and the "Iron Curtain" across Europe.

From Berlin, I flew home to the States to visit with Mom, Dad and Jack in St. Louis and then with Aunt Peg and Uncle Marian Halsey in Tulsa. At the end of my leave, I flew to New York and then to Lisbon,

Portugal, where I sampled their famous port wine. I also went to a bullfight, which was interesting because in Portugal they padded the horns of the bull and did not kill it (as they did in Spain). From Lisbon I flew to Rome, Italy, and then back to Tripoli, Libya, where I arrived at the end of July 1959.

After a month in Tripoli drawing maps and filling out paperwork, I was assigned to the first wildcat exploratory well in our Concession 59, located deep in the desert in eastern Libya. This well was designated the A1-59 (first well in Concession 59) and called "Waha" from a local landmark. The well was located in a desolate and little explored area about 150 miles south of the Mediterranean Sea, which I had visited on a reconnaissance trip back in December 1957.

On one flight to the rig our plane was engulfed in a dust storm and the pilot lost his bearings. We circled around in the blinding dust and found the seacoast and then headed back in the direction of the wellsite. We got lost again and so our pilot called the nearby British Air Force Base at El Adem, famous as an air base during the desert campaign against German General Rommel during World War II. El Adem advised our pilot to land at their field and wait until the dust storm subsided, which we did. We spent the night in a barracks and had dinner and breakfast in their mess hall. The next day the skies were clear and we found the rig without problem. On this well, I started smuggling a bottle of port wine to our alcohol-free wellsite when I returned from field leaves. After a 12-15 hour shift on duty in the geological trailer, a small glass of port secretly enjoyed in my bedroom was a real treat.

We cored and drill stem tested several excellent zones of oil and when the well was completed in late 1959 it was proclaimed the fourth giant oil field found by Oasis. We immediately planned another well in the Concession 59 area.

SAND DUNES IN CONCESSION 71.
This picture shows a colleague of mine out in the sand dunes
near our drilling location in Concession 71. We enjoyed walking
in the dunes to watch the sunset as the air cooled off.

It was obvious from the good oil discovery at the Waha well that we were going to drill a number of wells in our Concession 59. But, this area was over 500 miles southeast of Tripoli, which made it difficult and expensive to fly people and supplies to the area with our twin-engine C-47 aircraft. It was decided to move our supply base for Concession 59 from Tripoli across the country to Benghazi, Libya, which was Libya's second capital city. This shortened our truck and airplane routes to Concession 59 to about 300 miles.

In November 1959, it was also decided to move the wellsite geologists involved over to Benghazi, and that included me. I packed up my household goods and clothes and the company DC-3 flew them to Benghazi. And, I wanted my car in Benghazi in order to go skin-diving and to once again tour around the Greek and Roman ruins in the Province of Cyrenaica, Libya. So, I picked up another geologist as a passenger and we started east in my MG-Model A sports car along the coastal road for the drive of 620 miles around the Gulf of Sirte and across the sandy and desolate central part of the country to Benghazi.

On the way, we stopped in a café in Misurata before crossing the desert part of the trip. In the café, a well-dressed Libyan came over and asked if we were driving to Benghazi. We said we were and he asked if he could "drive in convoy" with us; he was obviously terrified at the thought of crossing the desert alone. We begrudgingly agreed. Then, we left the café and were driving east at high speed in my sports car when I noticed that this Libyan was driving right off my left rear fender. We came to a stop and I angrily told this man that what he was doing was very dangerous and to fall back well behind our car but where we could still see him. He reluctantly agreed and we drove the rest of the way to Benghazi without incident.

I moved into the Oasis Oil bachelor's quarters in Benghazi. This was a nice large apartment overlooking Benghazi harbor. When in town, the other bachelors and I ate most of our meals at the old Berenice Hotel on the harbor, which I had visited back in May 1958, when I visited my old friend Jim Knudstad in Tolmeita. The Berenice served mostly English and Italian cooking that was quite good. And, in the basement was a seedy nightclub, referred to as "the snake pit," which provided a bar and live music nightly.

For recreation we drove to the spectacular Greek and Roman ruins at Tolmeita, Cyrene and Apollonia, all of which I had previously visited in 1958 and knew pretty well. We also enjoyed snorkeling in a large, fresh-water sinkhole just up the coast called "the Blue Lagoon." This deep circular hole was filled with fresh water eels about two feet long and we enjoyed swimming amongst them.

During my time in Benghazi the city was frequently engulfed in storms of red dust blowing out of the Sahara. Visibility went to nil while these dust storms blew. Soon after my arrival in Benghazi, I was assigned as the only geologist on a wildcat exploration well located deep in the Sahara Desert about 500 miles southeast of Tripoli and about 300 miles straight south of Benghazi in the Libyan Province of Cyrenaica. As this was the second well to be drilled in our newly awarded Petroleum Concession Number 59, the well was designated the "B1-59" and given the name "Defa."

The drilling rig was trucked across Libya from Tripoli and then south through the sand dunes to the location and the tall derrick erected ready to drill. I flew to the site and prepared the equipment in the geologist's trailer to log the progress of the well. An exciting project lay ahead of me as no wells had ever been drilled on this large and remote geological structure and the oil potential was unknown.

I felt a large sense of responsibility, as this well, like the others I had been on, was the culmination of millions of dollars worth of geological and geophysical surveys and interpretation. If I missed a showing of oil or gas while the well was drilling, it could result in a loss of many millions of dollars of investment and future revenues for both the company and the country of Libya. Also, I was the only Oasis representative on the well, which was staffed and crewed by our drilling contractor. On top of this, short-wave communication with our Tripoli office was unreliable because we were so far away. So, I was faced with the possibility of having to make big decisions on my own. I was not worried, however, as I had already gained much experience at drilling in the Sahara.

Normally, we had two geologists on each wellsite working 12-hour shifts, but I was the only geologist on the B1-59 because Oasis Oil Company had a shortage of geologists due to the large number of wells drilling. We had already discovered world-class oil deposits in Libya closer to the Mediterranean Sea coast. Also, I was by this time one of the more senior wellsite geologists with Oasis and the logical choice to take charge of this important wildcat well. When we drilled in areas of known oil we often had a petroleum engineer at the site to test and gauge any oil or gas flows encountered but there was no engineer on the B1-59 because it was a wildcat well on an untested prospect.

As we drilled ahead, I sat in the geologist's trailer by myself around the clock, as much as possible, performing my duties, such as checking rock samples, watching the drilling rate and gas detector charts, and sending daily reports to our headquarters in Tripoli by short-wave radio. I entertained myself by listening to the short-wave broadcasts of the BBC and Voice of America, reading adventure novels, and taking photographs of the desert scenery. During this period I also continued drawing cartoons of the humorous side of life in a desert camp using a very fine-pointed, crow-quill pen and black drawing ink. And, then there was always the next adventurous vacation trip to plan.

Suddenly, as the rotating drill bit was drilling about one mile deep (over 5,000 feet), we got a fast drilling break followed by limestone rock cuttings saturated with oil staining and a "gas kick" on the gas monitor! I immediately shut down the drilling and reported to Tripoli about the oil show and told them I was going in to cut a 50-foot rock core. The crew pulled the drill pipe out of the hole and replaced the drilling bit with a diamond coring bit and barrel to cut a solid 50-foot-long rock column, which would show the rocks and oil in much better detail than the tiny rock cuttings.

At two or three o'clock in the morning the drill pipe was pulled out of the hole in 90-foot sections and stacked in the derrick. Finally, the core barrel emerged, the diamond-studded core bit was removed, and the core started to come out. Soon I had 50 feet of rock core steaming with hot drilling mud laid out on the derrick floor. I broke the core up into foot-long pieces and put them into wooden core boxes. The rig crew used rags to wipe the mud off the cores and we immediately noticed natural gas bubbling out of holes in the rocks. I slowly broke open the cores, smelled gas, saw good oil staining, and tasted them for salt water but noted none. Then I wrote out a detailed description of the core and took samples for examination under the microscope. After canning a few cores to be sent to our lab in Tripoli for detailed analysis, I took sections of core into my geology trailer where there was a full set of core testing equipment to get a quick analysis. The analysis looked very good.

I got on to the radio and called Tripoli to report the results of the core and to recommend that we run a drill-stem test (DST) to test the potential oil flow of the oil zone. Tripoli approved the DST so I instructed the rig crew to rig up the testing equipment. I examined the drilling rate log to find the top of the oil zone and told the crew to place the "packer" at this point. We waited for daylight so we could shut off all the electricity in the camp to avoid a fire hazard. The driller then eased off his brake and put weight on the DST tools at the bottom of the hole. The packer expanded and squeezed off the upper hole and the drill pipe was rotated to open the valve below the packer to allow whatever was in the oil zone to flow into the drill pipe.

On the rig floor, the driller attached a small hose to the drill pipe and put the other end into a bucket of water. The hose bubbled vigorously in the bucket and then my nose told me that we had gas to the surface; I recorded the time. I waited to see if the gas flow would increase, which would require me to set up an "orifice meter" at the end of a flow line to record pressures that I could use to calculate the gas flow in cubic feet per minute. I had done this on other oil drilling rigs, but in this case the flow was too small to require it.

After an hour or so the gas bubbles in the bucket slowed down and indicated that whatever was coming into the drill pipe had almost stopped. We decided to stop the test and see what we had recovered in the pipe. The driller and roughnecks started coming out of the hole and we could see gas fumes as they unscrewed each section of drill pipe. All of a sudden, we heard a roaring sound and black oil gushed out of the drill pipe and sprayed all over the rig, the trucks, the trailers and the desert. It was a short burst of oil that gushed as gas was released — just like shaking a bottle of Coca-Cola to make the Coke spray out. This short oil spray "unloaded" every time they brought another joint of pipe out of the hole. Then, as we got to the last several hundred feet of pipe, a solid stream of oil poured out on to the rig floor every time they unscrewed a joint. We counted the joints and recorded that we had recovered several hundred feet of crude oil in the bottom of the drill pipe. The black oil was a good grade.

Finally we got to the bottom of the pipe and the crew removed the bottom hole pressure gauge. I took the chart into my trailer and with a micrometer measured the various lines scratched on the black copper plate and converted them to pressure in pounds per square inch. The low pressures indicated that the oil zone we had just tested had a relatively low potential. Then, to complete my drill stem report I had to estimate the total amount of oil that we had recovered, including all the oil that had unloaded all over the camp. I got my clipboard with my test report sheet and went out on the rig floor. I gazed around at the oil sprayed all over the rig, trucks, trailers and the desert. Estimating the amount of this oil might have been a daunting task for an inexperienced geologist but as a Saharan veteran I simply looked around and said to myself, "Well, that looks like about 25 barrels." And, this wild guess is what I recorded.

I radioed Tripoli with the test results and told them it looked like we were still in the oil zone and recommended that we cut another 50-foot core and test again if the core looked good. Our Tripoli headquarters concurred and so we went back into the hole with the core barrel.

To make this long story a little shorter, I cored and tested a series of fifty-foot oil zones, which meant we had a very thick oil zone. The drill stem tests each flowed some gas and recovered relatively small amounts of oil. These were relatively small flows by Libyan standards (I had been on other wells where we had oil and gas flowing at high pressure into the pits with potentials of 25,000 barrels of oil per day).

In any case, I consider the B1-59 as my most exciting wellsite experience because I was the only company representative at the well. At 28 years old, I made all the decisions and did all geological and engineering work as described above. The oil zone that I had worked my way through on this wildcat discovery well established a new oil field in Libya. After many more and better wells were drilled, it became a world-class giant oilfield (the Defa Field).

14 ✴

King Tut's Tomb

In the fall of 1959 a letter arrived on my desk in Tripoli from my old high-school friend Jim Knudstad, an archaeologist. Jim was in Luxor, Egypt, on assignment with the Oriental Institute of the University of Chicago, and he invited me to visit him. He promised me a trip to King Tut's Tomb. Jim's invitation sounded like a splendid adventure, and I immediately decided to take a week's vacation to go see him. A month later, I boarded a well-used airplane owned by Misrair, the Egyptian state airline, and flew east across the Gulf of Sirte to Libya's other capital city, Benghazi, then over Tobruk and El Alamein to Cairo, Egypt.

The next day I enjoyed wandering through the busy and noisy markets, or *souks*, of Cairo, especially as I was fairly fluent in Arabic by this time and could bargain and joke with the merchants in their own tongue as we sipped cups of tea. Cairo is Africa's largest city where East and West blend together. Native Egyptians were first on the site, followed by the Romans some two thousand years ago. Then the Arabs arrived about 641 AD and in 969 AD built a new walled city called El Qahira (the Victorious), known in the West as Cairo.

Following instructions from Jim's letter, I made my way to a city square and caught a local bus to the Giza Plateau located just southwest of Cairo. As the bus climbed onto the plateau, I got my first exciting glimpse of the three great pyramids and the Sphinx. All these famous funerary monuments were built during Egypt's Old Kingdom from 2800 to 2350 BC. I joined a tour group to explore the largest pyramid, the Great Pyramid of Cheops. With our guide, we walked around the pyramid and then went inside and up the steeply inclined main passageway to a large burial chamber. The chamber contained only a granite sarcophagus with the top removed and empty inside. After the tour, I climbed up the large limestone blocks all the way to the top of the 481-foot-tall Great Pyramid, the world's largest stone structure. From the top I had a grand view of Cairo, the Nile River Valley, the other pyramids, and the Sphinx.

Next stop was the Great Sphinx. With the body of a lion and head of a king and carved out of solid rock, it is impressive, but I was disappointed to find it much smaller than I had imagined. It is only 66 feet high. It seems all the photographs I had ever seen were taken at ground level looking up at the Sphinx with the pyramids in the background; this makes the Sphinx look huge compared with the pyramids in the distance. Finally, I agreed to have the customary tourist photograph taken of me sitting on a camel in front of the pyramids but I was very wary of the camel herders—they get you up on the camel and then won't let you down until you pay an exorbitant fee. Then I caught the bus back into Cairo.

In the afternoon, I toured Cairo by bus. First we visited the mud-brick Step Pyramid at Saqqara, just south of Cairo, built as a tomb for King Djoser in 2630 BC. Next stop was the ancient city of Memphis, capital of Egypt's Old Kingdom and famous for its 80-ton alabaster sphinx. Then it was shopping in the Cairo bazaar, followed by a visit to the large and elaborate Mosque of Mohammed Ali. Our last stop was the Cairo Museum, which houses Egypt's archaeological treasures. There were racks of mummy cases, large statues, and glass cases full of all kinds of ancient jewelry and other precious objects. One room held the mummies of the royal kings and queens laid out side by side with the heads unwrapped (and in various states of decay). The main

VISITING THE PYRAMIDS AND THE SPHINX.
The ancient Egyptian monuments of the Giza Plateau just outside Cairo
were fascinating. And, I also had the customary tourist camel ride.

exhibit of the museum was the golden treasure from the tomb of King Tutankhamen (Tut), featuring the gold mask from his mummy.

The next evening I boarded a train with air-conditioned sleeping cars for the 400-mile trip south along the Nile River to Luxor. For twelve hours I shared a very comfortable compartment with an elderly Egyptian who said his prayers on the floor as we left Cairo and again very early the next morning. The train pulled in to Luxor station and my friend, Jim Knudstad, was on the platform to meet me. We had a Coca-Cola in the busy city square next to the station and discussed what we would do during my visit.

My archaeological tour got off to a fast start with a short drive from the railroad station to the Temple of Luxor, built by Amenhotep III (1417-1379 BC). Jim pointed out many items of interest and showed me the famous grand avenue lined with sphinxes. We then drove a few miles north to the Great Temple of Amun at Karnak, a huge complex of temples, obelisks, and columns built about 1000 BC. From Karnak we drove back towards Luxor to Chicago House, Jim's base of operations. Chicago House is the research center and living quarters of the Oriental Institute of the University of Chicago. This large, well equipped, although not air-conditioned, facility was built with a million dollars from John D. Rockefeller following the discovery of King Tut's Tomb in 1922.

Jim showed me to my guest bedroom, and then we enjoyed a communal dinner with the staff. I met Dr. Hughes, leader of the expedition, and was very happy to see Charles and Myrtle Nims again, having met them in Tolmeita, Libya, in 1958. The Nims had lived and worked at Chicago House for almost twenty years and I was especially impressed to find that Charles Nims' specialty was reading and writing Egyptian hieroglyphics, as well as being the expedition photographer.

The next day Jim and I walked into Luxor and rented bicycles for the day. We then boarded a small, crowded, open-decked ferryboat to cross the quarter-mile-wide Nile River. As we motored west across the river, we watched small cargo and fishing boats with tall lateen sails called *fellucas*, tacking up and down river. At the other side of the river we arrived at the most famous burial ground in the world, the Necropolis (city of the dead) of Thebes. Here are located the Valleys of the Tombs of the Kings and Queens, Tombs of the Nobles, and many large, ancient mortuary temples. During the period 2000 to 1000 BC, Egyptian royalty and officials were buried in splendor here in rock-cut tombs situated in cliffs and valleys high and dry above the Nile River's flood plain.

We cycled down the mile-long entrance road to the Necropolis passing by two 65-foot-high seated statues of King Amenhotep III, known as the Colossi of Memnon, and arrived at the Temple of Medinet Habu, built by Ramesses III about 1175 BC. At Medinet Habu, the Oriental Institute had been mapping and recording hieroglyphics for almost ten years and had another five years to go. Archaeology obviously requires a lot of patience!

At Medinet Habu, Jim and I watched Dr. Hughes as he sat at the base of a 50-foot-high wall checking recorded hieroglyphics that covered the entire temple in rows three feet high. Then Dr. Nims led us down a long flight of steps around several corners into a subterranean chamber where an artist was copying wall paintings using natural sunlight reflected from the surface by tall mirrors held by native workmen stationed at each bend in the corridor. This was really fascinating to watch.

JIM KNUDSTAD AND I IN FRONT OF THE COLOSSI OF MEMNON.
We stopped here while bicycling to the Necropolis (city of the dead) of
Thebes across the Nile River from Luxor, Egypt. The two battered statues
are 65-foot-high seated figures of King Amenhotep III (1417-1379 BC).

Leaving Medinet Habu, Jim and I pedaled about two miles to Deir
El Bahri, the majestic temple of Queen Hatschepsut built about 1500 BC.
After exploring the Queen's temple, we pumped our bikes up a steeply
inclined and winding road about a mile to the Valley of the Tombs of
the Kings. Finally we reached our main objective, King Tut's Tomb. The
tomb of the boy-king Tutankhamen, who died at the age of eighteen in
1352 BC, was discovered by British archaeologists Howard Carter and
Lord Carnarvon in 1922. It was found crammed full of ancient treasures,
including the famous gold mask and case that covered Tut's mummy. I
had just seen all these treasures in the Cairo Museum a few days earlier.

Tut's Tomb was closed to the public, as Mondays were the day off
for the tourist guides, but a guard was on duty. Jim showed the armed
guard his archaeological pass and he said we could enter the tomb but
that the electricity was turned off. Using our flashlights, carried just for
this purpose, we descended a flight of steep stairs into the dark tomb.
We passed through a doorway and down a short corridor into a small
antechamber with walls covered with paintings and hieroglyphics.
We then walked through another doorway into the surprisingly small

burial chamber. Our flashlights found a large, golden-yellow, quartzite sarcophagus covered with hieroglyphics. The lid was open and inside our flashlights revealed King Tut's coffin of pure gold weighing some three hundred pounds. It was quite an eerie adventure to explore King Tut's Tomb all by ourselves using flashlights. I just hoped we would not be subject to the mummy's curse that had supposedly killed a number of archaeologists who originally discovered the tomb!

The hour was getting late, so we bicycled back downhill some four miles to the ferry landing. After a short wait, the ferry took us back across the river to Luxor town, where we returned our bicycles and walked a mile or so back to Chicago House. We were very ready for dinner that night after bicycling about ten miles around the Necropolis of Thebes.

The next day we again rented bicycles in Luxor and took the ferry across the Nile River to Thebes. We bicycled past the Colossi of Memnon to the Ramesseum, which was the mortuary temple of King Ramesses II (1304-1237 BC). It covers an area the size of two football fields. We walked around as Jim pointed out interesting points of archaeology, and then we sat on a granite column and ate our sack lunches amidst a massive jumble of stone blocks and columns. The Egyptian winter weather was wonderful—sunny and warm but not hot, with clear blue skies. Leaving the Ramesseum, Jim and I biked up a hill to the Tombs of the Nobles. Here are the less elaborate tombs of the wealthy people who helped the pharaohs rule their empires. We visited the tombs of Nakht (an astronomer), Menna (a royal scribe), and Ramose (a high executive officer). At Ramose, we descended into an underground chamber and viewed the vizier's mummy, still in its original place.

The next day Jim had a different kind of adventure in store for me. I was to go along for some new archaeological exploration in the Valley of the Queens. Jim and I joined Dr. Hughes and several other Chicago House staff on the Oriental Institute's motorboat to cross the Nile River. Two of the Institute's old Chevrolet cars awaited us on the other side. We drove over to the Institute's project at Medinet Habu to get a ladder and then proceeded to the Valley of the Queens. Someone had spotted a small cave in the side of a cliff and our little expedition was going to see if it contained any antiquities. As Jim and I watched, the 50-foot extension ladder was placed against the cliff and a woman archaeologist, who had found the cave, climbed up and into the hole. A few minutes later she re-appeared with the disappointing news that the cave was empty.

Jim and I then drove one of the cars to the Tombs of the Nobles. Jim wanted to show me the tomb of Kheruef, where he was working underground by himself making architectural drawings. We arrived at the tomb of Kheruef, who was a royal scribe to Queen Tiy, the mother of King Akhnaton. This tomb, built in 1320 BC, had never been opened to the public and was deserted and locked up when we arrived. Jim produced a key and opened the gate into the site. We scrambled through a narrow opening in the ground. With our flashlights illuminating the darkness we slid down a steep incline of loose rock and landed in a dark, underground corridor inside the tomb. As we walked down the long corridor, Jim pointed out with his flashlight a large number of partially unwrapped mummies lying on the floor. He explained that common people often put their mummified relatives in nobles' tombs to be preserved by the high terrain and dry climate, and for religious reasons.

We continued along the dark underground corridor stepping over large rocks and mummified bodies. Jim showed me the long, white, curved gypsum crystals growing out of the ceiling. A couple of bats flew over our heads, which gave me a scare. The corridor led into a large, underground open court with columns. We went through a doorway, down another corridor, around a right-angle turn and then came to the door of the burial chamber. With his flashlight, Jim showed me partially burnt bundles of reeds on the floor in front of the door to the burial chamber that he said had been used as torches by tomb robbers thousands of years ago. Jim said the robbers had stolen most items from the tomb, but some remaining items had been moved to the Cairo Museum. Entering the burial chamber, Jim showed me a mummy on the floor, but he said it was not the nobleman Kheruef, for whom the tomb had been built.

The next day I took the train back to Cairo and flew back to Tripoli by way of Benghazi. It had been an exciting archaeological adventure in ancient Egypt; a place filled with strange ideas of imagination, religion, and mythology all mixed together.

15 ✴

Tripoli Office Duty

Upon return from my adventure in Luxor, Egypt, I was assigned to wellsite duties on the B2-59 and B3-59 wells to help define the extent of the oilfield we had found on our B1-59 Defa discovery well. Now the work started getting routine and not nearly as interesting as the rank wildcat wells I had been sitting on.

I had greatly enjoyed my three years in the Sahara Desert of Libya, which I call my "Lawrence of Arabia Period," but by 1960 I had had enough sand, sun and flies. I began to resent the fact that several geologists who had arrived in Libya long after I had were now working in the Tripoli office, while I was still sitting on wells in the desert. It was not my nature to complain or make demands on my supervisors, but finally I decided I had to tell management I wanted to be reassigned to an office job in Tripoli. After three years of pioneering work in the desert I felt I had earned this promotion.

I went to see our chief geologist, Bruce Shade, and told him why I thought I should be promoted to a position on his staff in the Tripoli office. Shade listened to my story and then said, "Fred, I didn't know you wanted an office assignment. I thought you liked it in the desert. If you want to work in the Tripoli office that is fine with me. I will come up with a Tripoli assignment for you right away because you have certainly earned it."

In a few weeks, Bruce brought me into Tripoli on his staff as "exploitation geologist." By this time, Oasis had some 15 rigs drilling in the desert, which was a huge operation, and 10 of these rigs were drilling production wells in already discovered oil fields. I was assigned the job of looking after the 10 development drilling rigs; an amazing number of rigs for one geologist to look after. I made geological and petroleum maps of all the oil fields in Libya, followed field drilling by the other oil companies, and helped plan where our next development wells would be drilled. I also took morning radio reports from, and gave instructions to, our wellsite geologists in the desert, regarding such matters as: what depths to cut cores, when to take drill stem tests, and when to run survey electric logs. It was a fascinating and challenging job that I enjoyed very much.

Now that I was living full time in Tripoli, I settled into life in the city. It was a pleasant time to be in Tripoli and there were many amenities for American oilmen. At this time (many years before Colonel Gaddafi took over in 1969), living in Tripoli was much like living in southern Italy or Sicily because Italians owned most modern facilities. The other Oasis Oil Company bachelors and I enjoyed all the Italian-owned amenities, such as: restaurants, cinemas, casinos, nightclubs, sidewalk cafes, bars, beach resorts, jewelers, clothing stores, and grocery stores. There were also local nightclubs with belly dancers and native musical instruments. In these days, we oilmen were very welcome in Libya and there was no worry at all about roaming around to all of these places, both day and night.

We bachelors snorkeled and spearfished on nice beaches west of town and enjoyed sumptuous dinners at the two local luxury hotels. Occasionally, we visited the spectacular ruins of the old Roman cities of Leptis Magna and Sabratha some miles out of town. We frequently spent the entire afternoon at a sidewalk café drinking Beck's beer and eating pistachio nuts. It was a pleasant life for a 30-year-old bachelor.

I played golf three or four days a week at the Sea Breeze Golf Club at the U.S. Wheelus Air Force Base, located just east of Tripoli. The course was mostly hardpan ground with rows of date-palm trees (that left bristles sticking out of our golf balls if we hit them) and oiled-sand greens. Oasis Oil had a company challenge ladder so we always had a foursome ready to play. I was not a very good golfer and usually played near the bottom of the ladder with a couple of the other Oasis bachelors. We were so bad that one of our colleagues came up with a new name for a score of "ten over par" that we called a "gorilla." But, over time my game improved and I moved on up towards the middle of the ladder. My best score then, and ever, was a 90.

As a member of the Underwater Explorers Club, I took lessons and became certified as a scuba diver by the owner, a crusty, old "Brit" left over from World War II, and bought a used Siebe-Gorman aqualung set from him. It was great fun cruising around underwater and shooting grouper and octopus with my speargun. When I shot an octopus (they were two to three feet long), I would clean it and then beat it on the rocks to tenderize it. As they were still pretty tough and rubbery, I would then grind them up to make "octopus burgers" on the grill, which the other bachelors enjoyed.

Every time I visited the Underwater Explorers Club, I had a little underwater game that I played. I had discovered a large hole in the rocks in about fifteen feet of water near the club and found that a very large grouper fish, possibly in the 100-pound range, lived in a cave in the hole. So, with spear gun at the ready, I would swim along the bottom, sneak up to the hole and then look over the side for the grouper. Well, that grouper didn't get that large for no reason, because I would shoot and always miss as it swam quickly back into its cave. This game went on for all the years I was in Tripoli.

I especially enjoyed the British Armed Services Yacht Club located in the picturesque harbor of Tripoli. I had taken sailing lessons and earned my certificate to sail their fleet of "GP14" sailboats. The GP14 was a 14-foot-long, plywood sailing dinghy with a large mainsail and a smaller jib in the front. It could carry four people but would really fly with a crew of one or two in a good wind. The Club had regular Sunday regatta races, and served an excellent Indian curry lunch, but I usually preferred to take out a GP14 single-handed during the week. In a stiff breeze with sails fully unfurled it was tricky to keep the boat from overturning, but it was fast and exciting sailing.

Back in 1958, I had made the mistake of buying a locally made sailboat with a very tall mast and clapboard hull. It leaked and had to be constantly bailed out; and, while anchored at the Club in a heavy storm it sank. But, my boat did lead to one day of high adventure in 1962. I decided to sail out of the harbor and then west down the coast to a popular beach about seven miles away by boat. Then, I would have to tack back and forth upwind to get back to Tripoli Harbor. I knew my boat did not tack upwind very well so I allowed five times longer to get back than it would take to get to my objective.

I rigged up my boat and made sure I had my coastal nautical chart and a hand-held compass. Since I expected to be back in several hours, I did not bother to tell the Libyan boatman my plans as I pushed off at about two o'clock in the afternoon. It took almost an hour to sail up-wind out of the harbor and then I steered down-wind along the coast west of Tripoli. It was a beautiful sunny day with a good wind at my back and I enjoyed the ride as I took sightings on the shore with my compass and marked my positions on my chart.

At about 5:00 p.m., I reached the popular "Kilometer Eight Beach" about five miles west of Tripoli and, without landing, "came about" for the up-wind trip back to the harbor. It did not take me long to realize that I was in trouble. First, the boat was slowly leaking and filling with water but I couldn't bail it out because I had to keep my hands on the tiller and ropes to keep the boat pointed into the stiff breeze. Secondly, I began to realize that each time I tacked out to sea and then back towards the shore, I was only gaining a little over half a mile back towards the harbor. After hours of tacking back and forth it started to get dark and I was still quite a way from the harbor, which was in sight up-wind some distance away. As darkness fell, I decided to make one last, very long tack out to sea that would hopefully allow me to tack back into the harbor entrance.

At about 8:30 p.m., I started my long tack out to sea in the dark. The boat was sloshing around with about six inches of water inside. However, I was not really worried because the harbor wall was in sight about half a mile away and I could swim to it if I had to. Then, as I was preparing to "come about" and head towards the harbor the breeze completely died and I was becalmed. So, I decided to row with my one oar over to the distant harbor wall. Progress was slow with water sloshing around in the boat but the stars were beautiful. As I rowed, the oar stirred up bright, shimmering phosphorescence in the water, just like I had read about in my adventure books.

Rowing towards the harbor, I saw a light on a fishing boat nearby and decided to row over to it. Upon arrival, I asked the crew in Arabic for a tow into the harbor and told them I would pay them well for it. They threw me a rope and towed me back into the harbor using their motor.

I was surprised, however, that instead of towing me back to the Yacht Club they towed me to the Harbor Patrol Police Station! Apparently several foreign sailors had been arrested a few weeks earlier in a boat trying to escape their ship and, therefore, all fishermen were told if they saw any foreigners in boats to report them to the police. Fortunately, the boatman from the Yacht Club saw us arrive and explained to the police who I was. I finally got home at about midnight after quite an adventure at sea.

From time to time, I would organize bachelor trips to the mountains south of Tripoli known as the Jebel Nefusa; these mountains are the tribal territory of the Berbers, who are blue-eyed descendants of native Libyan and Roman inter-marriages. We would drive our convoy of sports cars, including my MG-A, south from Tripoli into the desert and pass through the tiny outpost of Azizzia, which is world famous as having the highest temperature ever recorded on earth (136 degrees Fahrenheit). Then, after climbing a very winding escarpment over a thousand feet high, we would be up on the top of the green and pleasant mountains. We would drive around looking at Roman ruins and long abandoned, underground houses dug in deep pits (people who live underground are called "troglodytes").

Our favorite village in the mountains was Garian, which had a nice hotel with a bar and restaurant, and a nearby, very interesting camel market. While in Garian, we always made a trip to the ruins of a World War II field hospital just out of town to see the famous "Ladies of Garian." In 1943, Clifford Saber, a volunteer American ambulance driver with the British Eighth Army created three expertly drawn murals of nude ladies on the inside walls of the building in red paint (some said it was blood). The most famous mural covered a large wall and portrayed an enormous naked woman lying on her side, American pin-up style. The upper outline of her body was in the shape of the coastline of North Africa with towns and cities labeled; and all over were humorous little figures of soldiers, paratroopers, bombs and airplanes.

THE LADY OF GARIAN.
This mural was drawn on the wall of a field hospital near Garian, in the mountains south of Tripoli, in 1943 during World War II. The artist, Clifford Saber, was an American ambulance driver with the British army. The upper outline of the reclining figure is in the shape of the North African coastline with cities and towns labeled. The entire mural is dotted with small drawings of paratroopers, bombs, airplanes, etc. and also covered with graffiti.

My housing in Tripoli started out in 1957 with a shared bachelor apartment in downtown Tripoli. Several years later, I moved into a villa on the outskirts of town with three other bachelors, and finally I moved into my own penthouse apartment on the top of a bank building overlooking the King's Palace. My neighbor was the Oasis Oil Company resident manager, so I had "moved on up" to very nice quarters. As 1961 approached, I began to make plans for my semi-annual "home leave" trip back to the States. I enjoyed the thought that I could use my company air ticket to go out and come back through any European countries that I wished.

In March 1961, our recently formed Petroleum Exploration Society of Libya had its first weekend field trip outside Libya. And, it was to be a very exciting event. We were off to the Republic of Chad located on Libya's southern border in central North Africa. Chad had been part of French Equatorial Africa but gained independence in late 1960. But, as we headed south, Chad was still quite French and still retained its old, French city names (all later changed to local names). Also, the French Foreign Legion was still on patrol.

On Friday March 17, 1961, some 80 petroleum geologists, including most of my colleagues at Oasis Oil, assembled at Tripoli Airport and we took off in a chartered French DC-6B. We headed due south across the Sahara Desert of Libya. We passed over the great Murzuk Sand Sea and then looming ahead out of the haze we saw the dark peaks of the Tibesti Mountains, located in northern Chad.

We flew over the rugged Tibesti terrain composed of light colored ancient igneous rocks with pitch-black, extinct volcanic cones poking up here and there. Then we got a real thrill. Our French pilot banked steeply and then dropped our plane down into an awe-inspiring volcanic crater called *Trau au Natron*. This crater, on one of the highest peaks in the Tibestis (about 11,000 feet), is five miles in diameter and 3,000 feet deep. It was amazing to be looking out of our airplane windows up at the rocks as we circled inside the crater! In the center of the crater were snow-white deposits of sodium carbonate with small black volcanic cones sticking up through them. Flying on south, we circled over Lake Chad, a shallow lake dotted with barren rocky islands, and then landed at Fort Lamy (now renamed N'Djamena), the capital of Chad. We boarded buses and drove into town as our cameras snapped photos of men dressed in white flowing robes and bare-breasted women with loads on their heads. The people in this country are black and mostly Moslems.

We drove to a French Foreign Legion airbase and settled our gear into a barracks for the night. We each had a bunk bed with overhead mosquito netting. Then, it was off into town for dinner at a hotel. After dinner we found all kinds of peddlers assembled outside the hotel waiting for us. Laid out on the ground for sale were spearheads, knives, brass animals, python-skin folio cases and other crafts. I bought my share of everything. The next day we boarded our buses and went back into town to see the sights. We walked along the Chari River, watched fishermen in dugout canoes at work, and took photographs of all kinds of local dress, and undress, on the streets. The natives were very friendly.

Then we headed back to the airport and flew southeast along the River Chari to Chad's second largest city, Fort Archambault (now renamed Sarh). We drove to a hotel outside of town, had a picnic lunch and then were provided cots outdoors in the courtyard for the night. After lunch, we found a row of local vendors outside our hotel with crafts for sale, mostly ebony and ivory carvings. I bought an ivory crocodile about a foot long. We walked into town and visited the local market place. It was a colorful sight with women wearing bright-colored, printed dresses, and some with babies slung on their backs. All kinds of goods were for sale including food, baskets, china, glassware, and oil lamps. After shopping, we walked back to the hotel and had dinner.

The next morning we got on our buses and I was fortunate to get a seat on the top of a bus sitting in the luggage rack; great view! We slowly drove along a dirt road across dry, scrub country with scattered small trees. We enjoyed taking pictures of the locals walking along the road. Finally, we arrived at a primitive village of round, thatched huts with conical roofs. As our bus pulled into the entrance of an old, French farmhouse, the right fender demolished one side of the brick entrance. The locals swarmed around us with friendly smiles.

Our group walked into the center of the primitive and dusty village, which we then discovered was home to a tribe of Ubangi people. The most famous feature of the Ubangis is that the women wear wooden disks from three to eight inches in diameter inserted into their upper and lower lips. The explanation for this strange custom is said to date back to the days when this tribe disfigured their women this way so that Arab slave traders would not take them. The few Ubangi women with lip disks that we saw and photographed appeared to be quite elderly and we were told the custom was dying out.

Next, we all gathered in a circle and several men did a spear dance for us accompanied by native music on drums and xylophones. The dancers brandished eight-foot-long spears and would charge at us and then stop with the points of their spears quivering inches from our faces.

UBANGI WOMEN WITH WOODEN DISKS INSERTED IN THEIR LIPS.
We saw these old women in a dusty village in the southern part of the
Republic of Chad in Central Africa.

After the dancing, we went through the village buying everything in sight. We bought baskets, woodcarvings and other crafts, and some of the guys even went into the huts and bought wooden chairs right out from under the inhabitants. After our buying frenzy, we got on our buses and drove back to Fort Archambault airport. As we were waiting to board our plane, vendors came speeding up on bicycles and offered us spearheads and knives of very shiny steel that had obviously just been made the night before.

We flew 1,700 miles north back to Tripoli, where we arrived late Sunday evening after an exciting weekend in the wilds of Central Africa.

16 ✳

Around the World

In late 1960 I had begun to plan my second home leave trip from Libya back to the States. Oasis Oil Company's policy provided a first-class air ticket with overhead sleeping berth back to the U.S. every two years. Although the first jet-powered commercial aircraft were now in service, it was my good fortune that Oasis's policy was still based on the 10,000-mile round-trip being made by slow propeller-type planes, which at 260 miles per hour took almost 12 hours just to cross the Atlantic Ocean.

I decided to use the cash equivalent of a first-class ticket directly home to purchase a round-the-world ticket in coach class. This opened up the exciting prospects of visiting the Taj Mahal in India, seeing the temples of Bangkok, watching Chinese *junks* sailing around the harbor in Hong Kong, walking the streets of Tokyo, and sunning on the beaches of Hawaii. Most, but not all, of these dreams were to come true.

On April 7, 1961, I departed Tripoli late in the evening on an Egyptian Misrair flight east to Benghazi, Libya. We landed on an airstrip lined with kerosene lamps at Benina Airport in Benghazi. After a brief stop, our flight continued along the North African coast and landed in Cairo, Egypt, at two o'clock in the morning. I took a bus into town and checked into the well-known Semiramis Hotel. The next morning I walked around town and visited the colorful local markets, or *souks*, just as I had done on my previous visit to Cairo. My walk led me down to the bank of the Nile River near my hotel. I watched the *fellucas* tack their cargoes up and down the river lowering their tall, single lateen sail each time they passed under one of Cairo's bridges.

Then I spotted a small sailing dinghy tied up to the bank next to some rowboats. An Egyptian dressed in traditional turban and robe sat near the boats waving off flies with his flywhisk. I asked him in Arabic if I could rent his sailing dinghy, and he replied with a wide smile that I certainly could. We agreed on an hourly rental price and then I looked over the boat with an experienced sailor's eye. The dinghy was about 12 feet long with a red hull, white deck, and patched red mainsail and front jib sheet. I gave the boat a push away from the bank and jumped aboard. I lowered the centerboard, set the sails, and steered out into the Nile in light winds. This was an adventure I had not anticipated.

Out in the Nile, I slowly sailed up to several rowboats crowded with young Egyptian men all laughing, splashing their oars, and having a good time. As soon as they saw me approaching in my little sailboat, they knew I was a foreigner and were surprised to see me out in the middle of the river. They waved and shouted greetings accompanied by much laughter. I shouted back greetings in Arabic as I sailed by. After slowly circling around a couple of large *fellucas* and exchanging greetings with the crews, I headed back to the boat jetty. It had been an exhilarating and enjoyable two hours of sailing on the ancient Nile River.

That evening, I checked out of my hotel and boarded the bus for the airport in the dark. Suddenly a tire on the bus went flat. I sat patiently inside while the driver and a couple of native passengers got out the jack and started working on the front right tire. After 15 or 20 minutes I heard the Arabic getting louder and more frantic, so I got out to see what was going on. I grabbed a flashlight from the driver and was distressed to find he had been jacking up the fender of the bus instead of the axle. He had run the jack up so high that it had now jammed. I had dealt with this kind of mess for several years in the Libyan Desert, so I immediately took charge of the operation. I started giving orders in Arabic to dig the sand out from under the jack. Fortunately for me, a BOAC (British Overseas Airline Corporation) crew bus came by and stopped to see what was going on. I abandoned the airport bus and rode with a flight crew to the airport. At the airport, I joined this same crew on board a BOAC Comet jet to fly some 3,000 miles from Cairo to New Delhi, India.

It was a thrill to take my first flight on a jet-powered airplane. The Comet, which was the first commercial jet in service, was very smooth, quiet and fast. We flew all night towards India. At 6:00 a.m. on April 9, our Comet landed in New Delhi. BOAC arranged a low cost hotel for me, and I boarded the BOAC bus for town. The bus dropped me off at a very modest little hotel, The Hotel Broadway. I checked in and a porter took me to my room. My hotel room was modest but clean. There was one window with the shade pulled all the way down. I walked over and raised the shade and was shocked to find myself looking directly down into a miserable, dirty little village crowded with naked children, goats, dogs and water buffalo! This was more adventure than I could stand.

I immediately called for a taxi, checked out, and headed for the best hotel in Delhi, the Ashoka Palace. What a change! The Ashoka Palace was a huge, palatial place surrounded by large gardens and well-manicured lawns. I negotiated a small suite with all meals for the rupee equivalent of 15 dollars a day. The next morning I lounged around the Ashoka Palace's circular swimming pool and was waited on by turbaned waiters. In the afternoon I walked through crowded, noisy streets into town. I went into a rather tattered movie theater and saw the movie "Ben Hur" starring Charleton Heston. During intermission I was amused by a slide that flashed on the screen saying, "Why kill for food? Eat at Veeraswami's vegetarian restaurant."

The next day I again walked into town and found myself in the middle of a political rally of some kind with yelling crowds, waving signs, and trucks with blaring loudspeakers. I kept walking until I reached a quiet neighborhood of large, elegant bungalows and nice gardens. According to my map, the prime minister of India lived somewhere in this upscale suburb. I walked back to my hotel and enjoyed dinner in the ornate dining room.

The following day I went for another swim in the hotel pool. In the afternoon I took a bus tour of New Delhi. Our tour first visited the Jantar Mantar astronomical observatory, which was built during the glorious Mogul Period. The Mogul Empire was founded by Moslem invaders from Persia beginning when King Babur conquered Delhi from the Hindus in 1526. This dynasty ruled India for more than 200 years under Babur, Akbar, and Shah Jahan. Most of the famous palaces, mosques, and tombs which I was about to visit in India, such as the Taj Mahal, were built by the Mogul emperors during the 16th, 17th, and 18th centuries. Our tour then visited the Indian government parliament and secretariat buildings, India Gate (the large arch through which British kings and queens rode in huge processions during the British rule, or Raj), Humayun's Tomb, the Quwat-Ul-Islam mosque with its 235-foot-high tower, and a Hindu temple.

On April 12, I left the hotel at 7:45 a.m. for a bus tour 125 miles south to the town of Agra and the world-famous Taj Mahal. The old bus bounced along on a rough road, and the weather became very hot as the sun rose high in the sky. The bus did not have air-conditioning and hot air and dust blew through the open windows. We stopped to watch a snake charmer by the side of the road playing a swaying reed flute as a cobra in a basket followed the flute back and forth. We then continued south.

The bus hit a really big pothole, and I flew up in the air. When I came back down onto the seat, I felt a stabbing pain in my rear. I put my hand down to investigate, and it came back covered with blood. It seemed that a razor-sharp spring had poked a hole in the seat and stabbed me. I used my handkerchief to stop the bleeding, and at the next rest stop I found a neat right-angled cut right through my trousers, shirt tail, underwear, and me.

We finally arrived at Agra just before noon and visited several large, ornate Mogul tombs and mausoleums built around 1605 AD. After lunch at the Imperial Hotel, we drove up to the main attraction, the Taj Mahal. The Taj Mahal was very impressive and everything I had hoped and expected it would be. Built in 1631-41 AD, it took 20,000 workers some 10 years to complete this tomb for Mumtaz Mahal, wife of Mogul Emperor Shah Jahan. The ornately carved and polished white marble gleamed in the sun and, as a geologist, I was amazed to find the marble walls inlaid with multi-colored, semi-precious stones in the shape of beautiful flowers. Our tour guide showed us the false and true tombs of Mumtaz and Shah Jahan and pointed out the huge, ornamental gates on three sides of the tomb and the River Jumna on the fourth side. We walked around the large surrounding fountains of white marble.

THE TAJ MAHAL AT AGRA, INDIA.
My visit to this spectacular tomb of the Mogul Emperor Shah Jahan
and his wife was everything I had expected it to be and more. I was
especially interested in the marble walls, which had inlaid designs
of flowers and other decorations in semi-precious stones.

Next our bus drove to the nearby Agra Fort, also located on the Jumna River and in sight of the Taj Mahal. This fort of red sandstone with a moat was built by Mogul Emperor Akhbar in 1565-75 AD. While touring the fort, I met a white, South African photographer and freelance journalist about my age named Frank Fishbeck. As our interests were much the same and he was also headed to Bangkok, we decided to travel together. The next day I checked out of the Ashoka

Palace and moved my luggage into Frank's room at the Hotel Metro until my departure for Bangkok late that night. After lunch Frank went on a tour and I took a motor-driven rickshaw to the Red Fort in New Delhi. The Red Fort was built by Shah Jahan in 1639-48 AD on the bank of the River Jumna as his royal palace. It is a huge red sandstone fort surrounded by high walls and a moat. I walked through the ornate buildings and small well-kept lawns and was especially impressed with the Pearl Mosque and its walls of jewel-inlaid white marble similar to the Taj Mahal.

I then took a taxi to Raj Ghat on the outskirts of Delhi, where the body of the great Indian leader for independence, Mahatma Gandhi, was cremated following his assassination in 1948. The site consisted of a concrete platform about 15 feet square. Three men were covering the platform with red, orange, and white flower blossoms. I hailed a motor rickshaw for a ride back to the Hotel Metro. The drive-chain broke on the way, and I had to wait on the side of very crowded and noisy street while the driver repaired it.

Frank Fishbeck joined me for dinner at a vegetarian restaurant. We then wandered around town killing time until my bus left for the airport for my flight to Bangkok. We ended up taking my luggage to the tavern in the Imperial Hotel and watching a floorshow with a French singer. At midnight Frank returned to his hotel, and I continued waiting for my airport bus.

Very early the next morning I boarded an Air France Boeing 707 for the 1,800-mile trip to Bangkok, Thailand, where I arrived at 10 a.m. I checked into the Hotel Princess, which I had heard was favored by Pan American Airlines crews, and slept all afternoon to catch up on my night of lost sleep. On awakening, I asked my hotel concierge where to go for a drink and dinner. On his advice, I went to Chiquita's Bar for a drink and there I met a Marine guard from the American Embassy, who told me about other places to go for food and entertainment. I had an excellent dinner of barbecued spare ribs at Club Keynote and then went to Club Flamingo for an after-dinner drink and watched a French singer and combo perform. As I walked home to my hotel, I passed several very smelly, open sewers and was reminded of the amusing story I had heard in Tripoli of an American oil-field hand who had gotten drunk in Bangkok and fell into one of the open sewers.

On Sunday, April 15, I took a tour of the many highly ornate Buddhist temples of Bangkok. There was an entire city block of golden temples of all shapes and sizes with tall spires and fierce warrior statues. The tour then went across the Chao Phraya River to a temple decorated with pieces of broken pottery. From the temple, I gazed across the river at Bangkok and its gleaming temples just as the sun set behind me. That evening I took a night tour of Bangkok, which included a Siamese boxing match where the fighters used boxing gloves but also kicked with their feet. Then our tour went on to various nightclubs where we saw classic Balinese dancing, folk dancing, and a nightclub act in which scantily clad young ladies ended up spraying shaving cream all over each other.

I was up early the next morning to take a boat trip upriver to see the famous "floating markets" of Bangkok's canals. It was fascinating to cruise through crowded narrow canals lined with tiny shops with boats loaded with food and wares tied up alongside. Everywhere I looked were baskets of fish, flowers, fruits, coconuts, vegetables, clothing, canned goods, and hardware. The canal people wore colorful sarongs and straw hats and carried straw and blue silk parasols. We passed a house with smiling adults and children sitting, swimming, washing, and playing in the canals. It was a very exciting and lively sight.

That evening, Frank Fishbeck, whom I had met in New Delhi, arrived in Bangkok after a brief visit to Rangoon, Burma. We got together for dinner and then walked for several hours around Bangkok seeing the sights. The next day I hired a car and driver to visit a nearby village called Nakhon Pathom, which my guidebook said had the largest standing statue of Buddha in the world. Frank decided to join me but could only chip in a little money towards expenses as he was traveling on very limited finances. At 9:30 a.m. we headed west out of Bangkok for the 35-mile trip to Nakhon Pathom. We drove past farms and rice fields. People in straw hats were catching something in the ditches along the road using straw baskets turned upside down. As we crossed the Chao Phraya River, we stopped to visit some more floating markets.

We arrived at Nakhon Pathom, which my guidebook said had been continuously occupied since 150 AD. Our driver took us to see the largest standing Buddha, or *chedi*, in the world. This Buddha had been carved out of a solid rock wall in 500 AD and stood an impressive 380 feet high, or 75 feet higher than the Statue of Liberty. After viewing the Buddha, Frank and I ate some beans and rice cooked with coconut milk at a little open-air restaurant constructed of bamboo. Walking back to the car, Frank spotted a woman selling sweet "sticky rice" packed in bamboo tubes. He said he wanted to photograph the woman and eat some of the rice for a story he was writing. Against my better judgement, he talked me into eating some also.

We drove back to Bangkok and I agreed to meet Frank later that evening for dinner before I left for Hong Kong the next day. I took a nap and then went over to Frank's hotel. I waited in the lobby but Frank did not show up. Finally I asked the concierge if he had seen Mr. Fishbeck. The concierge told me Mr. Fishbeck had left a message for me saying he was sick and could not make dinner. I wondered if Frank had discovered a more interesting way to spend the evening than having dinner with me but, about an hour later, I knew he was telling the truth because I started to get sick. Very sick. I was up all that night with intense pain, vomiting, and diarrhea. No doubt that "sticky rice" in Nakhon Pathom had poisoned both Frank and me.

The next morning, I was still feeling very sick but had plane reservations for Hong Kong. I reluctantly boarded a Pan-American Airways plane for the 1,000-mile flight. Some hours later our plane descended, dodged mountains in thick clouds, and then emerged under the clouds to land on an airfield built out into a small bay surrounded by mountains. It was a long taxi ride from Kao Tak Airport into Kowloon, the Chinese mainland side of Hong Kong. I checked into the Hotel Miramar and collapsed into bed still very sick. In the morning, I was feeling a little better, so I walked around the town to do some shopping in dark, cloudy, rainy weather. The next day I was feeling even better and did more shopping at the fancy shops jammed full of all kinds of luxury items at good prices. At one shop I bought an 18-karat gold Omega Seamaster wristwatch for $196.50. The price of gold was fixed worldwide at $35 per ounce in those days while it is something like $300-400 an ounce today. I am still using that watch and the gold in it is worth more today than the watch cost in 1961.

I joined a boat tour across the harbor from Kowloon, on the Chinese mainland, over to British-owned Hong Kong Island. The harbor of very blue water was jammed with every kind of boat and ship imaginable. *Sampans, junks,* freighters, and yachts were traveling all different directions at every speed. On Hong Kong Island we took a bus up into the Aberdeen highlands to see the beautiful homes there with a grand view of the harbor and Kowloon. Back on our tour boat, I rode on the bridge next to the British captain so I could ask him questions. We cruised over to a large area with thousands of Chinese *junks* anchored side by side. The captain said all the boats were full of permanent refugees from China. He asked me to guess how many junks were there. I thought a minute and threw out a guess of "75,000." He said, "You're right!"

After the tour, I took a taxi to one of Hong Kong's famous "floating restaurants." Although I still didn't feel too well, it was very nice. The restaurant was a large boat tied up at a jetty in the harbor. The boat was gaily lit with Chinese lanterns, and the sides of the boat were lined with nets in the water filled with live fish waiting to be served up very fresh.

Since I was still not feeling totally well and had been traveling for two solid weeks, I decided to cancel my visits to Tokyo and Honolulu and go directly home to my family in St. Louis as quickly as possible. On April 21, I flew 1,850 miles from Hong Kong to Tokyo Airport. From Tokyo it was 3,500 miles to Honolulu, passing over Pearl Harbor just before landing. From there I flew 2,750 miles to San Francisco, where I stopped overnight at a hotel near the airport. From San Francisco I flew to Lambert Field in St. Louis, Missouri, where my family awaited my arrival.

I had a very nice two weeks with Mother, Dad, and Jack in Clayton and did my shopping for another year in Libya. I also made a quick trip to Tulsa to see Uncle Marian and Aunt Peg Halsey, who treated me with their usual fine hospitality. For my return trip to Libya, I flew from St. Louis to Boston, then to Shannon Airport, Ireland, and over to London, where I stayed at the Strand Palace Hotel, the same hotel my father had stayed in as a lieutenant in the Canadian army during World War I. From London I flew over the European Alps and across the blue Mediterranean Sea until the white, sandy coast of North Africa appeared on the horizon. I then arrived back in Tripoli, Libya. The total distance that I had covered circumnavigating the globe was approximately 22,000 miles.

17 ✴

Marathon Battlefield, Greece

On March 28, 1962, I arrived in Athens, Greece, from Tripoli, Libya. As an Ohio Oil Co. geologist on loan to the Oasis Oil Co. of Libya, I was passing through Greece on my way to the United States on home leave. This leave marked the completion of my fifth year in Libya with Oasis Oil. With the coming of faster, jet powered aircraft, Oasis Oil now provided annual home leaves to the U.S. in economy class, rather than the old plan of leaves every two years by first class on slow, propeller planes (12 hours just to cross the Atlantic Ocean).

I settled into my hotel in Athens and then decided to rent a car and drive to Marathon, Greece. During the eight years I had been with the Ohio Oil Co., the company trademark of the Greek Marathon runner, Pheidippides, and the company trade name, "Marathon," had always intrigued me. Here was a chance for me to see where it all originated.

At noon I went to a car rental agency where I was informed that my Libyan driver's license (in Arabic) was not valid in Greece and that I would need an international driver's license. The clerk advised that the office that issued such licenses would close in 30 minutes, but that I could get one if I hurried. First I had to run over to a nearby park where a photographer with an ancient-looking camera on a tripod took a photograph of me and developed four prints in less than 10 minutes in tanks mounted on the tripod legs. The car rental agency then sent a man with my passport and the photos to get my new license, and he was successful.

At 2:30 p.m., an almost new Volkswagen was delivered to my hotel and I departed for Marathon in warm, sunny weather. After a pleasant drive through mountainous farmlands east of Athens, the highway turned northward along the hilly shore of the Aegean Sea, and I soon arrived at the town of Marathon, about 22 miles northeast of Athens. Marathon is a small farming town in a mountain valley by a small river. The farmers in this region produce olive oil, cotton, and a variety of fruits and vegetables.

The first thing I had intended to do upon my arrival was to photograph the road sign at the entrance to the village, but to my dismay a truck had torn the sign down and it was flat on the ground. Proceeding into the village, I looked for some sign of tourist facilities where English was spoken but found none. I did, however, find a group of Greek men sitting in the sun in front of a filling station (the original "Marathon" gas station?) and decided to wade in and try to make myself understood. I walked over, pulled up a chair, and sat down in their midst. They looked puzzled and must have wondered what I was going to do next.

I then took my English-Greek phrasebook from my pocket and pointed to the Greek equivalent of "I want an English-speaking guide." There was a good deal of laughter and smiles, but they shook their heads in the negative. Next I showed them the Greek phrase for "I am interested in archaeology." Again there was much discussion, and finally one man motioned for me to drive south and he would accompany me. We drove south of town to a farm where several people were working in a field. My new friend waved to the farmers, and a man ran over to our car.

The middle-aged farmer who joined us spoke very good English and said his name was Anastassios Chryssinas. Anastassios had learned English while attached to NATO (North Atlantic Treaty Organization) forces in Turkey some years earlier. He volunteered to act as my guide for the afternoon. Driving a little farther south of Marathon town, we came out of the hills onto the Plain of Marathon, which lies on the shore of the blue Aegean Sea, surrounded by mountains and marshes. It was on this plain that the historic Battle of Marathon was fought in 490 BC. In the center of the battlefield is a 30-foot-high conical dirt mound, which is said to cover the remains of the Greek warriors who fell in the battle. Climbing to the platform on the top of this mound, I had an excellent view of the battlefield.

As I stood on this ancient mound, I remembered that my high school hero, Richard Halliburton, had stood at this same place some 35 years earlier. In his 1927 book, *The Glorious Adventure*, Halliburton proclaimed that the battle fought here had "settled the cultural destiny of our Western World." My guide, Anastassios, explained how the famous battle was fought. The landing by ship of a Persian army 60,000 strong at Marathon in 490 BC constituted the first direct Persian attack on the Greek mainland. A Greek army of 11,000 marched from Athens to repel the invaders. The Greeks fortified the hills surrounding the Plain of Marathon and boxed the Persians in against the beach. Then, the Greek *hoplites*, heavily armed with helmets, shields and spears, formed into deep ranks and files (phalanxes) and marched on the Persians that had landed on the beach. Seeing that the Greeks were very fierce and had superior position, most of the Persians withdrew to their ships and sailed away, leaving a rear guard of 20,000 men to meet the Greeks and cover their retreat. The Greek army, under the leadership of Miltiades, charged the remaining Persians and by a flanking movement surrounded and defeated them with over 6,000 Persians killed.

According to Anastassios, the Greeks still tell a story about a Greek warrior who ran to the beach and took hold of a retreating Persian boat with one of his arms. The Persians cut off his arm, but he grabbed the boat with his other arm, which they also cut off. In desperation the poor fellow grabbed the boat with his teeth, so they say, but then the Persians cut off his head! When the Battle of Marathon was won, an Olympic runner by the name of Pheidippides was sent to Athens to carry the good news. Pheidippides ran 22 miles and 1,470 yards to Athens, where he fell dead of exhaustion after gasping with his last breath, "Rejoice; victory is ours." The Battle of Marathon was decisive because it dealt the Persians such a blow that they did not attempt another attack for 10 years, thus giving the Greeks time to prepare their army. It also gave the Greeks wonderful encouragement because the Persians had been greatly feared and were considered almost invincible.

According to Anastassios, the German army, while occupying Greece during World War II, re-staged the Battle of Marathon with 3,000 German troops dressed in ancient Greek and Persian battle-dress. The show was for the benefit of some high-ranking generals from Berlin, who viewed the scene from atop the burial mound where we now stood.

THE BURIAL MOUND AT THE MARATHON BATTLEFIELD IN GREECE.
This is the burial mound of the Greek soldiers that died defeating the Persian army at Marathon in 490 BC. During World War II, German troops occupied Greece and they staged a reenactment of the famous battle, which was viewed by German generals standing on this mound.

The victory run of Pheidippides to Athens is celebrated by Marathon foot races, which have been held all over the world since 1896. The Ohio Oil Co. trademark of a Greek runner wearing a loincloth and the trade name "Marathon" also commemorate the run of Pheidippides to Athens. (Later in 1962, the company name was officially changed to the Marathon Oil Company and the Greek runner trademark was dropped after a Madison Avenue firm's research found that people thought it was advertising men's underwear!)

After touring the battlefield, Anastassios and I drove back to the town of Marathon. At the edge of town we stopped and propped up the road sign, which read "Marathon" in both Greek and English. Anastassios took several photos of me next to the sign to record my visit. We then drove to Anastassios's home near the town church, which he said his grandfather had helped build when he was mayor of the town. We sat down in the parlor and were served coffee and cake by the maid. While we were talking, Anastassios's wife and several neighbor women and their daughters came by and joined us for coffee, but none of them spoke English. Anastassios translated the conversation for me and explained to me that it was considered polite among friends to dunk the rather hard cake in the coffee before eating it.

PROPPING UP THE SIGN AT THE ENTRANCE
TO MARATHON, GREECE.
This photograph of me holding up the town sign appeared
in the *Marathon Messenger* of September 1962.

It was getting late in the day, and so it was time to say goodbye and leave these very hospitable people. As I left the house, Mrs. Chryssinas picked a flower and gave it to me. Using a sketch map that Anastassios had drawn for me, I took a mountainous, but shorter, road back towards Athens. After traveling for an hour or so, I came to Lake Marathon, the source of water for the city of Athens. It was a pretty little lake nestled in the mountains with an impressive dam. The sun was setting in the mountains behind the lake as I arrived. After driving across the dam, I joined a main highway and drove southwest back to my hotel in Athens.

From Athens, I flew back to the States where I spent a very sad and disturbing two weeks with my family in Clayton. Dad had had a severe stroke some months earlier and I found him in a hospital bed in our small apartment. He did not respond mentally and had lost control of most of his body. Mother and Jack had to feed and care for him around the clock. Jack remained cheerful as he was devoted to Dad, but it was taking a terrible toll on Mother.

As I had one more year to serve in Libya, about all I could do for Mother was assure her that Uncle Marian and I would certainly continue our financial support. I told Mother if she needed anything to let me know. After two weeks in St. Louis, I flew to Tulsa, Oklahoma, to see my Aunt Peg and Uncle Marian Halsey and stay at their home, which they jokingly called "the Halsey Hilton" because of their many visitors. They were doing fine and I had a nice visit with them.

I flew back to St. Louis to buy clothes and other things for another year in Libya. Then, I said goodbye to Mother, Dad, and Jack and flew back to Libya with a heavy heart because of their situation. This was to be the last time I would ever see my father as he died suddenly just a few months later from another stroke before I could make it back from Libya.

18 ✳

A Paradigm Shift
and Departure from Libya

In 1962, I could look back on four years of adventurous and pleasant bachelor life in Tripoli and Benghazi, Libya. This life had included camping in the Sahara Desert, taking part in major oil discoveries, exploring fabulous Roman and Greek ruins, Oktoberfests in Munich, sailing, scuba diving, golf, tennis, trips all over the world and much more. However, sometime back in 1961, my thinking had begun to take a radical change. I began to feel like the world was passing me by and it was time to think about getting married and settling down to family life. I guess my thoughts were much like those of American soldiers who were isolated overseas for years during World War II.

And so, at age 30, I decided to abandon my bachelor lifestyle and to start dating with a long-range objective of marriage. However, I soon found out that I was a good example of those men who devote their lives to themselves and their work and then suddenly decide it is time to think about getting married. I lacked many social skills when it came to this department.

There were not many single, western women in Tripoli in 1961. I started dating American and Italian oil company secretaries, American schoolteachers, and American embassy staff. I enjoyed this new life, but also found myself making awkward and embarrassing social blunders. This was rather distressing for me, as I had always tried to do everything very well, but I persisted.

Eventually, I met an American girl and we found we had many interests in common. We started steady dating and enjoyed each other's company. Some months later, however, I began to realize that she was seriously thinking about marriage. I did not feel ready to get married yet, and it soon became evident to her that marriage would not work for us. Needless to say, our relationship ended and this was quite distressing to both of us.

This break-up was the first time in my life that I had gotten into a situation that was beyond my control, affected the lives of others, and was embarrassing. I felt upset and my self-confidence was shaken. And, to make things worse, I was living alone in a foreign land with nobody to talk to about my problems. Fortunately, I had interesting and challenging work at the office to help get me through this troubled period.

I soon realized, however, that I needed to talk to someone about my problems and so I went to see the chaplain at the U.S. Wheelus Air Force Base, located just outside of Tripoli. No doubt, my boyhood Sunday school classes led me to this action, although as a boy I thought the classes were a waste. I talked things over with Chaplain McQueen in his church office at Wheelus. He helped put my situation in perspective and said I would just have to put everything behind me and move on. He asked if I would like to help him prepare for his Sunday services for a few weeks and I readily agreed. He also gave me a Bible, which I still have and consult from time to time.

As a result of my paradigm shift in thinking, I began to examine my attitude toward people in general. I began to realize that while my Uncle Marian in Tulsa had made a successful geologist out of me and given me much self-confidence, he had also loaded me down with some negative "baggage." Although a generous man who could be quite charming, Uncle Marian did not like certain groups of people, especially blacks and Democrats. And, he had little patience, and was frequently very harsh with anyone who displeased him for any reason.

I began to realize that I had picked up some of Uncle Marian's short-tempered behavior. I had frequently been harsh with the drilling crews in the desert and now was often impatient with some of the men I worked with in my office. I recalled that one year, when I had been Oasis's chief geologist as a summer replacement, our temporary exploration manager said to me as he was leaving Tripoli, "Fred, you are a good manager with a promising career, but you are too hard on your staff. You need to let up on them a little." It became apparent to me that I had problems in this area, too.

As is my procedure when I have problems, I started reading every book I could find on human psychology and social interaction. I learned that people pattern their actions, reactions, and protective barriers based on their childhood experiences. If a person feels that they have been hurt, controlled, or embarrassed in childhood they can develop quite a "chip on their shoulder" and harbor various angers deep inside. Then, years later if a traumatic event occurs it can trigger all kind of emotional responses such as anger and depression, but it can also wake a person up to what they have been doing. As one book I read put it, it is like you are riding sleepily along on a train through life when all of sudden a suitcase falls out of the overhead rack, hits you on the head and wakes you up.

My books also made it clear that it is not easy to make a fundamental change in your attitude toward people even when you know you need to do it. One book said that making fundamental and painful life changes is like plunging into a very wide river and then battling your way across against the currents and obstacles until you finally reach the other side as a new person—if you have not given up.

As a result of my self-doubts and determination to change, I "jumped into the river and swam against the current," and tried to change my life. I felt like giving up many times, and retreating back into bachelor life, but slowly began to be more thoughtful, friendly and open with other people. I also realized that I had been so wrapped up in my own life and problems that I had not been paying as much attention as I should have to my family's problems in St. Louis.

Meanwhile, in early 1962, Marathon Oil Company (as the Ohio Oil Company had been renamed) sent a representative from Findlay, Ohio, to Tripoli to interview all Marathon employees who were on loan to The Oasis Oil Company of Libya, such as me. The reason for these interviews was that Oasis was about to become a separate company, jointly owned by Marathon, Conoco and Amerada, just to operate its prolific oil production in Libya, which was now over 100,000 barrels of oil per day (later it would reach one million barrels per day). During my interview, I was given a choice of returning to Marathon at the end of my current Libyan contract or staying on permanently with Oasis Oil with enhanced pay and benefits.

At about this time, some geologist working on a well in the desert posted the following lament on the office bulletin board:

WHO ME?

I'm drunk, sick and hungry, and got a hangover, damn flat broke, need a shave and a haircut, homesick, lonesome, tired, and no mail for a month; missed the plane to town, no Libyan exit visa, no friends and damn few relatives. In debt, poor character rating, inefficient, pay all screwed up, lousy chow, no clothes, laundry lost, vacation plans rejected by the boss, lost my inoculation record, lost my passport, lost my Libyan identification card; been working night shifts for two weeks, camp generator shot, no air conditioning, and diesel oil in our drinking water. Missed the morning radio report to Tripoli and the chief geologist wants to talk to me after I finish testing this well. Got a "Dear John" letter from home, thirsty, sleepy, just about to poop in my pants and the damn latrine is off limits until after the company VIPs leave from their camp inspection, and then some SON OF A BITCH says to me
SIGN OVER TO OASIS OIL FOR THE BENEFITS!!!!!!!!!!!!!

As I had been in Libya for almost five years and was getting real tired of the place, I had no hesitation in deciding to return to Marathon Oil in the States in mid-1963 after almost six years of service in Libya.

My work as the Oasis exploitation geologist continued to be very interesting and challenging as the company now had 13 drilling rigs operating in Libya and most of them were drilling development production wells that fell under my responsibility. Our Sidrah oil production pipeline from Concession 32 to our Es Sider marine loading terminal was also about to be extended south to include our large new oil discoveries in Concession 59, which would greatly increase our total oil production.

Then, while I was going through my mental struggles of trying to reinvent myself, and also thinking about returning to the States, I received a telegram from my Uncle Marian Halsey in Tulsa. The cable informed me that my father had died on December 9, 1962 and that funeral services would be held on December 11. It further said that all arrangements had been made and that everyone in the family realized that there would not be time for me to get home to attend the funeral. And so, on top of all my other disturbing issues, I now felt very sad about missing my father's funeral and worried about how Mother and brother, Jack, were doing.

I was also noticing with alarm that things were changing for the worse in Tripoli. What was once a quiet and relaxed city was becoming jammed with oilmen and oilfield vehicles as a result of many huge oil discoveries in the Libyan Desert. Local Libyan youths were getting very frustrated at all the foreign money and power pouring in and they started getting drunk in sidewalk cafés and shouting obscenities at passing foreign women. [As a result of this shocking local behavior in a Moslem country, I was not really surprised when Colonel Qaddafi took over the country in 1969 and kicked all the foreigners out, including the Italians.]

Even though my situation in Tripoli was getting very tiresome, my mind began to calm down and some months later I emerged with a happier and more considerate attitude toward people. I began casual dating again but I was more relaxed and enjoyed everything more.

In the spring of 1963, as my time in Libya neared its end, I was informed by Marathon that I was being assigned to the international exploration division at the company headquarters in Findlay, Ohio.

Now it was time for me to prepare to leave Tripoli after six long years. I started selling off my scuba diving equipment and other things I would not need back in the States. I closed bank accounts and prepared to ship my clothes and various memorabilia back to Findlay. But, in order to get an exit visa to leave Libya, I had to go around to various government offices and get their signatures on my exit visa application. For example, the police had to certify that I had no charges pending.

I arranged to sell my MG-A sports car to another foreigner living in Tripoli, but was then informed by the Libyan customs department that I would have to pay the original customs duties on the car before I could sell it and leave Libya. I was furious and went to our company's chief local representative and explained to him that I should not have to pay the customs as another foreigner with customs-exempt status was buying the car. We drove to the customs offices and he asked me to wait outside in the car. He came back some time later and said the problem had been solved. I wondered whom he had paid off and how much money had been involved. After this incident, I was really ready to get the heck out of Libya!

I finally got all the required signatures on my exit visa application, submitted it to the authorities and was granted permission to leave Libya. I packed up all my oil paintings, spears, knives, bows and arrows, camel saddles, photographs, fossils, binoculars, silver and brass work, ivory carvings, and other mementos of my years in Libya. My fellow Oasis bachelors, with whom I had spent so many adventurous years, threw a nice going away party for me at the Hotel Uaddan and invited the entire Oasis office staff and their spouses. The boys also gave me a very nice, Kangaroo-skin golf bag as a going away present.

Finally, on May 10, 1963, I said goodbye to my friends and my faithful Libyan houseboy, Ali, who was in tears, and a company car took me to Tripoli airport. I breathed a huge sigh of relief as I saw the yellow, sandy coastline of Libya disappear behind me into the haze as my plane headed north out over the blue Mediterranean Sea toward Rome, Italy, and then home.

And, thus ended my six years in Libya, North Africa. As I left, I carried with me many happy memories of exciting adventures. And, also some sad memories, too. But, I was comforted by the thought that I was leaving Libya at age 32 years a more mature and caring person than had arrived at age 26.

19 ✴

Findlay, Ohio

In May 1963, I made my final departure from Libya and after almost six years I was glad to leave. I had greatly enjoyed my adventurous years in Libya—especially my "Lawrence of Arabia Period"—but I was greatly relieved to leave the Libyan coastline behind me.

After brief visits to my old familiar haunts in Rome, London and New York, I arrived in Clayton, Missouri, for my first visit with Mother and Jack since Dad had died the previous December. It was quite sad to find Dad gone, but at least the terrible burden of his round-the-clock nursing care was lifted from Mother and Jack. I did what I could to make Mom and Jack more comfortable.

After leaving Clayton, I flew to Tulsa to visit with Aunt Peg and Uncle Marian Halsey. Uncle Marian was pretty well retired from his oil wildcatting days but had bought a very small and old oil field south of Tulsa that was being water-flooded by surrounding wells to force the last underground oil to the surface. Uncle Marian loved to drive out to his oilfield and look into the crude oil tanks and smell the oil. The project was not commercially viable and used up most of his life savings, but it kept him happy at age 67.

In June 1963, I flew to the Findlay, Ohio, headquarters of the Marathon Oil Company, to take up my first assignment in the States for six years. The "cultural shock" I was to experience in this small Ohio farm town was to rival many Third World countries I had visited.

I reported to work and was assigned as a regional geologist in International Exploration. I immediately started work on many interesting projects in Europe and the Middle East involving geological maps, reports and published data to try to find new places to drill for oil. And, more importantly, I learned how things were done and thought through, or sometimes not thought through, in the headquarters of Marathon International Oil Company; this experience was to prove very valuable for the rest of my career with Marathon.

Upon arrival in Findlay, I rented an apartment in the newly constructed Findlay Garden Apartments on South Main. I decorated it with new furniture and my memorabilia from Libya. Although I had been driving small MG sports cars for many years, I decided MGs were too small for my golf clubs and luggage and so I bought a used, but immaculate, Pontiac Bonneville convertible; it was very large with bright red exterior and white leather interior.

I did not know anyone in Findlay except for the geologists I worked with at Marathon and they were all family men. As a result, I decided to explore the region around Findlay, which is located in northwest Ohio about 40 miles south of Toledo. So I ventured forth to places like Detroit, Toledo, Cleveland and Cincinnati. I traveled alone to check out the sights, restaurants, symphonies, operas and golf tournaments. However, it did not take long to see everything of interest and I kept falling asleep on the long drives back to Findlay. I then concentrated on activities in Ohio towns nearer Findlay like Lima, Upper Sandusky and Bowling Green, but other than one or two nice restaurants, these small towns were of little interest.

For recreation in Findlay, I played golf with some geologists I had known in Libya, and briefly took up handball at the local YMCA until my wrists were so badly bruised I decided to give it up. Out of desperation, I even tried bowling in a company league for a brief period. At the large company offices, home base of Marathon Oil, I passed up the crowded, well-lit, chrome-plated, non-alcoholic company cafeteria, and instead always had lunch at a bar down the street, where most of the locals would not have been caught dead. But, I enjoyed the tasty lunches, good beer, and entertaining barmaid, who had sarcastic remarks for all her regular customers.

Then, I was invited to join Toastmasters International to learn how to deliver speeches and for the social activities. I joined up and attended one dinner meeting each week. The comradeship was pleasant and I learned much about public speaking, as I was required to give a speech of some kind at every meeting. [I used these speaking techniques in my presentations to company management for the rest of my career.] One big advantage of belonging to this group was that I got to know the top Findlay people from the mayor on down to major and minor businessmen.

To this day, whenever I think of small American towns like Findlay, I always recall the time in Findlay when a fellow Toastmaster asked me to come over to his house for dinner and to show he and his wife my slides from Libya. The three of us had a enjoyable evening, but the next morning I was shocked to see an article on the front page of the *Republican Courier* newspaper with the headline, "Marathon geologist gives lecture on travels in Libya." The story related that I had given a lecture illustrated with slides at the home of a local heating contractor. The article, which my fellow Toastmaster obviously had given to the newspaper as soon as I had left his house, failed to mention that there had only been three of us there! Anyway, I settled into a comfortable but rather quiet life in Findlay and work at the office was interesting.

In July 1963, I made a business trip to New York City that changed my life forever and much for the better. While in New York, I decided to drop by and say hello at the Marathon Oil Company office at Rockefeller Center. Upon my arrival, I was introduced to Marcia Mehl, who said she would be happy to show me around the office. Well, it was definitely a case of "love at first sight!" I was swept off my feet by Marcia's beauty, smile, charm, and friendly manner. I was thrilled when Marcia agreed to join me for dinner that night. Over dinner, Marcia told me she was from Findlay, Ohio. She said she had been working for Marathon in Findlay when she heard through friends of her parents that Marathon was expanding its New York office and she decided to leave Findlay, with its small population of 25,000 people, for the "Big Apple."

Marcia said she drove to New York City with her parents and talked to the manager of Marathon's New York office, Millard Saul. He told her she could come to work for them but would have to quit her job with Marathon in Findlay first and possibly lose her employee benefits. He said that no female employee had ever left Findlay to pursue work at another Marathon office. Marcia then quit her Findlay job and started work in Marathon's New York office at the beginning of July 1963.

The day after I had dinner with Marcia, I flew on the Marathon plane back to Findlay with Marcia Mehl very much on my mind. I soon found out that Marcia had very kindly alerted her girl friends in Findlay that a new bachelor was in town. All of a sudden, I was getting invitations to all kinds of parties and met all kinds of new friends. My social life really picked up in Findlay.

Marcia came back to Findlay for the Labor Day holiday, and we had lunch and watched the parade wind its way through the town. Marcia said she really liked my red Pontiac Bonneville convertible. After she left to fly back to New York, I knew that Marcia was someone very special.

Thanks to Marcia, I had a pleasant summer and fall in Findlay with parties and picnics with newly found friends, most of whom worked for Marathon. I also played golf and enjoyed my Toastmaster group.

In October, my boss in Findlay called me into his office and asked me if I would like to move to London, England, to start up an exploration office there. Findlay was beginning to get a bit boring, so I quickly said, "Yes." I then got excited about moving to London, except for having to sell my nice red convertible.

Marathon's tax department informed me that I needed to talk to Chase Manhattan Bank in New York about British tax procedures. I pondered over this banking matter and then thought of Marcia Mehl in New York. I told Marathon that I would need to fly to New York to personally talk to Chase Bank but did not mention that I also had in mind seeing Marcia again before I moved to London. A few days later I flew by company plane to La Guardia Airport in New York and checked into Marathon's suite in the Barclay Hotel. The next day I went to Chase Bank's offices and discussed British tax matters with them.

Then I headed straight to the Marathon Office in Rockefeller Plaza and asked Marcia to have dinner with me. She accepted my invitation and that evening I picked her up by taxi at the Barbizon Hotel for Women, where she was staying. On the way to pick her up, the taxi driver amused me by telling me that the Barbizon was a hotel for women, but there were lots of ways for men to sneak in from the back and spend the night with the residents.

Marcia and I had a wonderful evening at a restaurant called something like "Camelot" that had a drawbridge entrance and swords and armor hanging all over. The next day I flew back to Findlay to start getting ready for my move to London. Marathon's tax department told me that in order to earn foreign tax credits against my U.S. taxes for 1964, I would have to arrive in London on or before the last day of 1963 and thus be resident in the United Kingdom for the full 1964 tax year.

On November 22, 1963, I was packing my belongings and watching President John F. Kennedy's visit to Dallas, Texas, on the television. All of a sudden, the entire Kennedy assassination unfolded before my eyes. That is one date in history I do not have any problem remembering!

I finished packing up my bags for London and put my furniture in storage. I then sold my beautiful Pontiac Bonneville (which really hurt) and flew to London in mid-December 1963.

20 ✳

Assignment London

After arriving in London in December 1963 to take up my new assignment there as an exploration geologist, the first thing I did was to write to Marcia to tell her of my new surroundings. At this time, a new British band, called The Beatles, was the pop music sensation with such numbers as: *I Want To Hold Your Hand*, *A Hard Days Night* and *She Loves You*. I went to Carnaby Street and bought a tea towel with pictures of the Beatles, and mailed it to Marcia in New York.

My first home in London was a small service flat (apartment with maid service) in The White House, a large sprawling group of white buildings near Queen's Gate. I took the "underground," or subway train, from Queen's Gate to Bond Street and then walked to the Marathon office. The office at that time was in the penthouse suite on top of Berger House (home of Berger Paints) on the south side of Berkeley Square.

The manager of the London office was Al Lager, whom I had worked for in Libya several years previously. The entire staff consisted of only about a dozen people, mostly involved with chartering oil tankers to haul oil from our oilfields in Libya to refineries in Europe.

I was the first geologist assigned to the London office, but I was soon joined by another geologist, Bill Swales, who years later would become the president of Marathon Oil Company. Bill Swales and I ushered in the year 1964 at a New Year's Eve party at The Eve Club on Regents Street. The party concluded just after midnight with a conga line that went backstage through the dancers' dressing rooms. Our assignment in London had gotten off to a good start!

At this time, Marathon was setting up a refining and marketing organization in Geneva, Switzerland, to sell all the crude oil being produced in Libya. Thanks to the efforts of my colleagues and myself, we had found huge oil reserves in Libya, which would last for a very long time. Marathon was building a refinery at La Coruña, Spain, and a string of gasoline stations in Spain was being considered. An interest was bought in a German refinery and 90 gasoline stations were purchased in southern Germany. A petrochemical plant was being built near Munich, Germany, and a string of over 1,000 gasoline stations was bought in Italy. Marathon Oil Company was being transformed from a small U.S. oil company to a major international oil company. It was an exciting time to be working with Marathon.

In April 1964, I moved from The White House to a very nice building of luxury service flats at 55 Park Lane, located on Park Lane right next to the prestigious Dorchester Hotel and across the street from Hyde Park. My apartment, Flat 57, had a living room, a glassed-in dining room, one bedroom, a small bathroom and a closet-sized kitchen. In typical bachelor fashion, I decorated it with my African bow-and-arrow sets, daggers, spears and other memorabilia from Libya.

Everything was relatively cheap in London in those days and my service flat rented for only $450 a month, which Marathon paid for. But, when the word got to Findlay that a bachelor was living at 55 Park Lane the executives with Marathon International got upset. A few weeks later the president, vice president and exploration manager of Marathon International flew into London with their wives supposedly on an office inspection tour. While in town, they ordered me to set up drinks and hors d'oeuvres one evening for them so they could check my place out. I guess my small flat filled with old fashioned, over-stuffed furniture seemed all right with them, especially at the reasonable rent it was costing Marathon, because I never heard any more about it.

I spent my free time as a bachelor checking out old and famous pubs (e.g., Ye Olde Cheshire Cheese, Dirty Dick's Tavern, The Prospect of Whitby, The George Inn, and The Waterman's Arms), sightseeing and collecting beer mugs. I had lunch almost every day at The Audley Pub and Grill near Berkeley Square. I frequently dined in the evening as a member of the private Siegi's Club, 46 Charles Street, off Berkeley Square. Siegi's consisted of one large and dimly lit dining room with widely spaced tables around the walls and a piano player. I also often ate at The Guinea, 30 Bruton Place, also off Berkeley Square.

Meanwhile, Marcia in New York and I were having a "Trans-Atlantic romance" by exchange of letters.

At the office, Bill Swales and I were directed by Findlay to concentrate our oil exploration efforts on countries where Marathon had refineries and/or gasoline stations, namely Spain, Italy and Germany. Meanwhile, I became a Fellow of the prestigious Geological Society (of London), which is the oldest geological society in the world having been founded in 1807.

In early 1964 I was assigned the job of finding new oil deposits in Italy, and I started compiling geological and petroleum maps of Italy and Sicily. My studies showed that huge natural gas deposits had been found in the Po River Valley of northern Italy, but the only oil fields in the whole country were quite small and located in southern Sicily. On closer examination of my data, I discovered that a very small oil field had been developed in the 1930s about 50 miles southeast of Rome near a village called Frosinone (pronounced, "frooze-a-KNOWN-eh"). This provided a clue that larger oil deposits could be hiding in the same area.

I got approval to spend the spring and summer of 1964 doing a field reconnaissance geological survey in the Frosinone area. I flew to Rome, rented a roomy Fiat 2500 automobile and bought large scale Automobile Club of Italy roadmaps, which showed every tiny road and village in my proposed mapping area between Rome and Naples. I then drove southeast out of Rome on the main *autostrada*, or national highway, towards Naples.

After passing Frascati, home of one of Italy's best-known wines, I found myself driving in the narrow Liri River valley flanked on both sides by steep mountain ranges looming up about 5,000 feet high. These mountains are part of the Apennine Mountain chain that runs like a spine down the whole length of Italy. The road I was driving was originally built by the Romans and I could almost hear the booted tramp of ancient legionnaires.

I soon came to the little village of Frosinone located in the middle of the valley just off the *autostrada*. After asking directions to the old Ripi oil field, I found a small cluster of rusty, old oil pumps that were still pumping away. Total production at this field was less than 50 barrels a day, while in the international oil industry we look for tens or even hundreds of thousands of barrels a day from a single field to be commercially viable.

For my night's lodgings, I chose a small *pensione*, or bed-and-breakfast type hotel, in Frosinone, which I was to get to know quite well that summer of mapping. After dinner I studied my maps and decided the largest oil potential was located in the hard limestone rocks that formed the high mountains on each side of the valley, rather than in the younger and softer rocks in the valley.

Armed with my Brunton survey compass in a belt holster, a rock hammer, road maps, geologic maps, and a notebook, I started my survey the next day by driving east up into the mountains. I also had with me a very early model, battery-operated pocket tape recorder to make notes as I was driving along looking at the geology; the only problem was that as the batteries ran down, the tape slowed and the pitch of my voice recordings got lower and slower.

Driving slowly up the steep, winding roads, I stopped every now and then to examine a limestone rock outcrop, record the data in my notebook, and plot the location on my roadmap. Finally I reached the top of a rugged mountain range at the Italian ski resort of Filletino. The resort, surrounded by snow-capped peaks reaching 6,000 feet above sea level, was deserted as the 1963-64 ski season had already ended.

I worked my way south through the mountains and was pleased to find that the limestone outcrops on the mountain sides and road cuts were well exposed and inclined at relatively steep angles, which made my mapping easier. As I drove around every available road and track, I stopped frequently to measure the direction and amount of inclination of the rock layers and any breaks, or faults, that I observed and record them in my notebook. Between stops I recorded general geological observations on my pocket tape recorder. In my hotel room every evening, I plotted all the day's data on my Automobile Club maps and drew form-line contours to show how the rocks had been bent and broken by mountain-building forces. I was looking for a place where the rocks had been bent upward into a high spot, known as an "anticline," which is the usual place to find oil trapped deep under the ground.

After two weeks of mapping in the mountains, I returned to Rome and then flew back to London for two weeks, where I worked up my field notes into oil exploration maps for our management to examine. By this time, the North Sea oil and gas boom had started and several more geologists had arrived in the London office. Our office was moved to larger space in The Bechtel Building on Portman Square.

For the next six months I continued my pattern of two weeks in Italy followed by two weeks in London. I got to know Rome's streets, piazza's, fountains, and restaurants very well that summer, which was fun and exciting. I frequently visited the Roman Coliseum and Forum and could visualize the past activities of the ancient Romans there. I particularly enjoyed having coffee on the famous Via Veneto with its fashionable coffee bars and restaurants, and also enjoyed dining at a small family-style restaurant called *Sergio y Ada* near the Trevi Fountain.

As the months went by, I mapped the geology southward down the Apennine Mountains towards Naples. I worked my way through towns with names like Subiaco, Isernia, Fondi, Alatri, Sora, and Arpino. Many of these places were actually old, medieval, fortified towns perched on mountaintops surrounded by high walls. I came across several monasteries in the mountains, such as San Beneto, located southeast of Subiaco, and was occasionally surprised to find a small, fertile, green valley with farms and grazing horses.

The people I came across in these mountain areas did not speak any English and, as a result, my Italian language skills rapidly improved.

I learned enough Italian to do the essential things, such as service my car, order food, and stay in small hotels; in the oil field vernacular we would call this "bedroom and bar-room Italian." In some very small villages I would find the one local restaurant closed, but the family would open the doors and take the chairs off the tables just to serve me lunch or dinner. They usually asked me, in Italian, if I wanted spaghetti and beefsteak, as these were the easiest things for them to cook on short notice. So, I would sit there alone having my private dinner.

In some villages I got the distinct idea that they had never seen a foreigner before. One day I met an old mountain man walking along a narrow, rocky road. I greeted him in Italian, and he started talking so rapidly that I understood only a few words. After a long conversation, I managed to inform him in my rather poor Italian that I was an American. He was amazed and told me about all his relatives in the United States.

In the small hotels I stayed at up in the mountains, I would sometimes go into their little lounge after dinner to watch television, but all in Italian, of course. It was amusing, for example, to watch a movie where John Wayne would swagger into a saloon and boom, "Buon giorno!" There was no "night life" in these remote villages, and so after dinner I would retire to my room to examine my maps, read pocket books, and listen to my short-wave radio. By unwinding my 30-foot long antenna wire and throwing it out the window, I could listen to the British Broadcasting Corporation (BBC) in London and the Voice of America (VOA) in Washington, D.C. broadcasting in English. These links kept me company up in the wilds of Italy.

After months of mapping my way southward, I found myself looking up at Monte Cassino, the famous 6th Century Benedictine monastery located on a mountaintop some 1,700 feet above the city of Cassino. In 1944, towards the end of World War II, the German army made a bloody stand here against the American Fifth Army and British Eighth Army while retreating towards Rome. The town of Cassino at the foot of the cliff was totally destroyed; and then a completely new city built after the war a few miles away from the ruins. The monastery was badly damaged by allied bombers but later fully restored.

**EXAMINING MY GEOLOGIC MAPS AT THE
FOOT OF MONTE CASSINO, ITALY.**
Here I am on the balcony of my hotel room with the Monte
Cassino Monastery looking down on me. The monastery was badly
damaged during World War II, but later was fully restored.

At Cassino I had reached the southern end of the area I had chosen to map, and my fieldwork was done. I decided to reward myself by visiting the famous Roman ruins at Pompeii located on the side of Mount Vesuvius about 10 miles southeast of Naples. I drove south on the *autostrada* past the turn off for Naples and then up Mount Vesuvius.

Pompeii proved to be a fascinating archaeological site. What had been a favorite resort for wealthy Romans was completely demolished in 79 AD by an eruption of Mount Vesuvius that covered the city with 13 feet of volcanic ash. Excavations had revealed very well preserved public buildings, temples, theaters, shops, and private houses. Also found were perfect molds of some of the 2,000 people who were buried in the ash; by pouring plaster into the molds, complete human forms of the victims in various agonizing postures had been produced.

After a brief visit to the city of Naples, I started north up the coast road along the Tyrrhenian Sea back towards Rome. The next day I arrived at the port of Gaeta, home of the U.S. Navy Sixth Fleet. I spent the evening and night in a small hotel with a gorgeous view of the harbor. The next day I was delighted to find a local wine still being sold in clay amphora, just as the ancient Romans bottled it.

From Gaeta I proceeded north on an inland highway that follows the old Appian Way, which was built by the Romans through the Pontine Marshes. These marshes brought diseases to the Romans and were useless wetlands until Benito Mussolini reclaimed them in 1926 and put them into cultivation. I drove the old Appian Way through orchards and fields of artichokes and passed through the new cities of Terracina and Latina that Mussolini had created. I particularly remember with pleasure being served a huge plate of fried artichoke hearts at a local restaurant.

As I approached Rome from the south, I detoured over to the ancient Roman City of Ostia at the mouth of the Tiber River. Ostia was once Rome's principal seaport and naval base, and the ruins of the city are quite well preserved. I took a tour of the ruins, which reminded me of the splendid Roman ruins that I got to know so well in Libya. I arrived back in Rome and checked into a nice *pensione* with a charming view of the Trevi Fountain. I spent the next day lounging around all my favorite haunts in Rome, as I knew my fieldwork in Italy had come to an end and I might not be back in Rome for some time.

Back in London I worked up my Italian notes and maps, which indicated that I had located an oil-drilling prospect in the mountains east of Frosinone. In my report I recommended that Marathon Oil Company obtain an oil concession on the area and drill an exploratory well. After careful consideration Marathon's senior management decided that I had mapped a nice prospect, but the oil potential was too small to warrant expensive and difficult drilling operations in such mountainous terrain.

A few years later the Italian State Oil Company, AGIP, drilled a deep exploration well very near the prospect I had mapped, but it was a dry hole. Since the odds of finding commercial oil production in an unproven international area are 50 to 1 against it happening, I guess the result was not surprising. In any case I had spent an interesting and adventurous six months of geological mapping and getting to know the people and the sights of the region of Italy between Rome and Naples.

My reconnaissance geological survey technique in central Italy was a new mapping procedure for Marathon that I sort of invented as I went along. Upon completion, Marathon was impressed with the results and decided to have me do another such survey in Spain. They selected Spain because they had a refinery there, were planning a chain of gasoline service stations, and the first oil discovery in Spain had just been announced.

In June 1964, I started preparing a petroleum evaluation map of Spain. My mapping disclosed a very large and long anticlinal structure located southeast of Madrid in the dry basin country of the Tajo River. This anticline could have very large oil potential in a very strategic position if it checked out in the field. One day, I was requested to take my map over to the suite of Mr. John Donnell, President of Marathon International Oil Company at the Hotel Intercontinental on Park Lane. I ended up with my large map of Spain spread out on the floor of Mr. Donnell's hotel room. I got down on my hands and knees and pointed out the large anticline I had mapped. Donnell and his vice-president, Jim Anderson, were very interested and said an oil discovery there would fit their plans exactly.

I was instructed to immediately proceed to Madrid, rent a car, and check out the petroleum geology of the anticline near Tarancón. They also told me that a geologist who was being transferred from Tripoli, Libya, to London would join me. This turned out to be Frank Reilly, with whom I had worked in Ardmore, Oklahoma before I went to Libya. They also wanted me to teach Frank how I did my field reconnaissance surveys.

I flew to Rome to finish my business there and then flew to Madrid. I took a taxi to the Torre de Madrid (Madrid Tower) and there a porter carried my suitcase up to one of the Marathon Oil Company suites. Having worked with 640 Italian Lire to the dollar, I grossly over-tipped the porter. Then, Frank Reilly arrived from London, where the British Pound was worth $2.40, and he grossly under-tipped the same porter.

I took Frank to the offices of the Automobile Club of Spain and we bought very detailed, large-scale (1:100,000) road maps covering the Tajo Basin of central Spain and the Duero Basin of north central Spain.

Then, I explained to Frank the geological reconnaissance procedures that I had developed in Italy. I told him how I had evaluated the petroleum potential in Italy and even mapped an oil drilling prospect.

We rented a Seat car, which is an Italian Fiat assembled in Spain, and drove southeast out of Madrid across a rather flat and dry, dusty plain. We headed towards Tarancón to check out the anticline I had mapped in London. We arrived at the anticline in a few hours and were quite excited to be checking out a prospect that had captured the attention of our senior management in Findlay, Ohio. We found an excellent road cut, which passed right across the anticline exposing all the rock layers.

Driving very slowly, we passed though the road cut and noted the geology. We were immediately shocked to find all the rock layers standing in a vertical position! Rocks are deposited on ocean bottoms in a horizontal position and later are usually gently folded into up-and-down segments with the "up" parts having oil prospects. If the rocks are vertical it means that tremendous mountain building forces have squeezed and broken the rocks, which destroys the oil potential.

And so, unfortunately, our observations in the field indicated that the anticline I had mapped in London was really a highly disturbed, anticlinal fault zone with no oil potential. We immediately drove back to Madrid so I could telephone our manager in London with the disappointing news. I phoned our exploration manager in London, Hugh MacDonald, and he was quite disappointed by the news. He was not looking forward to reporting this to headquarters in Findlay, Ohio. After thinking for a minute or two, Hugh decided to fly down to Madrid and drive out to the anticline with us to confirm our findings before he called Findlay.

Hugh arrived the next day and we drove out to the structure. He looked at the geology and agreed with our evaluation. My big oil prospect did not prove out, but such disappointments are all part of the oil exploration game. Back in Madrid, we discussed the situation with MacDonald and it was decided we would continue our reconnaissance surveys around central Spain and then go up north to the region where oil had been recently discovered. Hugh went back to London to give Findlay the bad news.

Frank and I drove southeast from Madrid down the highway towards Valencia in flat dusty country with poor to nil rock exposures.

We spent our first night on the road in the town of Tarancón. We checked into a small hotel and were amazed to find the nightly rate was $1.45 per night per room! But, I soon found out that it was not worth even that, because my room was on an inner courtyard with a church bell at the top. The bell clanged every hour and half-hour all night long.

From Tarancón, we turned east off the main highway, left the dusty plain and started climbing up into some very pretty mountains. At dusk we arrived in the interesting little town of Cuenca. Cuenca is located on the western edge of some wooded mountains at an elevation of about 3,000 feet. The old part of the city lies on the edge of a steep and narrow valley (called a *cuenca* in Spanish) and it was interesting to see old houses projecting over the edge on wooden supports. The narrow, crooked streets were very picturesque, and the city famous as an artist's colony and a tourist spot.

Frank and I had some good *tapas* (snacks) at a bar followed by a tasty dinner. Then we went back to our little hotel, located right on the city square. I got into bed and was reading when all of a sudden a slightly out-of-tune, brass band struck up a military march right outside my window! I swung open my window shutters and was surprised to find the square alive with activity at ten o'clock at night. Children were running around, having had their *siesta*, and everyone was all dressed up to promenade around the square. I realized that I was not going to get any sleep for a few hours and so decided to go outside and join the activities. Frank had the same idea and joined me. We mixed with the local folk and then went back to our hotel rooms at about midnight. We were soon to find out that it was usual for everyone to have dinner around ten o'clock.

The next few days were spent driving around country roads south of Cuenca mapping the geology, which had weak indications of oil potential. Each night we headed back to Cuenca. Then, we drove south to the city of Albacete on the Madrid-Valencia highway. We spent the night in Albacete and found that it was famous for making knives. We visited several knife shops and bought a few at very cheap prices. From Albacete, we continued our surveys to the southwest into the region of La Mancha. La Mancha is famous as the land of Don Quixote and his companion Sancho Panza, characters from a novel written by the Spanish author Miguel de Cervantes in the 1600s. Just as expected, we found the rather barren landscape dotted by picturesque old windmills, just like the one that a delirious Don Quixote charged with his lance thinking it was an enemy knight.

ONE OF DON QUIXOTE'S WINDMILLS IN LA MANCHA, SPAIN.
One of several windmills that Frank and I drove past while doing geological reconnaissance in the La Mancha Region of central Spain.

We drove around La Mancha examining the geology and then found a small anticlinal structure that appeared interesting. But, after checking it out, we found it was only a very small local feature with no petroleum potential. By this time, my colleague, Frank Reilly, was really getting into this reconnaissance type of geological surveying.

After a few more days we decided that we had done all the surveys that we needed to do in central Spain and that it was time we traveled up north to the Duero River Valley near the recent oil discovery. We drove back into Madrid and our Madrid Office told us we could stay in their penthouse suite at the top of the Torre de Madrid, the highest building in the city. We checked into the suite and were very pleased with the luxurious accommodations and the spectacular view of Madrid. We decided that our marketing people in Madrid, who were planning a series of gas stations throughout Spain, lived a pretty good life. That evening, one of our marketing managers came up to the penthouse to have a drink with us and see how our surveys were going. He told us that Sophia Loren had once stayed in this suite.

The next day we drove northwest out of Madrid up onto a high plateau leading into the Sierra de Guadarrama Mountains. We passed a busy scene on our left with a large palace-like structure and found out it was a movie film company shooting scenes for the movie, "Doctor Zhivago." They were using white marble dust in place of snow.

Then, we reached the historic town of El Escorial at the foot of the mountains, where King Phillip II of Spain had built a castle and monastery in the 1500s. A few miles out of town, we arrived at the Valle de Los Caidos (Valley of the Fallen), a gigantic monument to the Spanish Civil War (1936-1939) built by General Francisco Franco, the military dictator who was running the country (but not very well). We saw a concrete cross nearly 500 feet high in front of a huge crypt tunneled out of solid granite inside a mountain. We went inside and joined the crowd to tour the huge project that was built with prisoners-of-war after the Civil War. In the back of the crypt was a large chamber that Franco had prepared for his own burial.

We continued our drive northwest and came down out of the mountains onto the plains of the Duero River Valley. This was another area that I had considered to have good oil potential for Marathon. We surveyed along to the city of Valladolid and then turned northeast towards Burgos, where a Gothic cathedral holds the remains of El Cid, Spain's 11th Century military hero.

After a couple of days, we concluded that the rock outcrops in this area looked very promising for oil exploration. Then, our surveys completed, we headed back towards Madrid. While Frank was driving, we came over the crest of a high mountain and had just started down the other side when the car's brakes went completely out! Frank had the presence of mind to steer the car off the road and sideswipe the rocks in the road-cut until the car came to a stop. Then, Frank managed to put the car into very low gear and slowly go down the mountain and creep into Madrid, where we returned our rented car. We decided the Spanish Seat had inferior workmanship to its Fiat origins; during our surveys screws kept falling out of the inside of the car and then the brakes failed!

In Madrid we again stayed in Marathon's penthouse suite, and spent a few days sightseeing. We saw the bullfights at the Plaza de Toros and visited the world famous Prado Art Museum, known for its collection of works by Valesquez, Goya and El Greco. We visited the crowded street "flea market" called El Rastro, where I bought a nice old percussion-cap pistol (stolen years later in Pakistan). I stopped at a jewelry store and had a solid, 18-karat gold band made for my gold Omega wristwatch that I had bought in Hong Kong some years earlier. The price of gold was still set worldwide at $65 an ounce, so my heavy, hand-made, gold band only cost $125.

Finally, our tour of Spain was over and we returned to London at the end of August 1964 to report on our surveys. We told our management that we thought the Duero River Basin around Burgos had the highest potential and recommended that a more thorough geological evaluation be made of that region. However, we were then informed by our executives that Marathon had decided not to build any service stations in Spain because the Spanish State Oil Company had a monopoly on too much of the gasoline market. They said that Marathon had lost interest in exploring for oil in Spain. But, anyway, Frank Reilly and I had had an enjoyable and adventurous month touring around Spain, or "eating our way across Spain" as Frank often referred to our trip.

I submitted my expense account for the Spanish trip to my supervisor, and for the first time experienced being chewed out over my hotel expenses. I was told, "Kelly, don't ever report a hotel room for $1.45 a night. It makes it look bad when the rest of us stay at expensive hotels in Paris and Rome. At least make it $10 a night!"

21 ✳

Bermuda and Wedding Bells

I had been sending letters and photographs to Marcia in New York all during my fieldwork in Italy. Now that the fieldwork was finished, it was time for me to plan my 1964 home leave back to the States. I kept thinking about Marcia in New York. Finally, in August 1964, I wrote to Marcia and suggested she meet me in Bermuda for a vacation together. She agreed and we worked on our travel plans by letter. I suggested three different hotels at three different price rates, which Marcia said she appreciated.

On September 4, I flew from London to Bermuda and was surprised that the flight was over five hours long. I found out that Bermuda is almost 3,500 miles from England but only some 800 miles from New York. Marcia's flight from New York was only two hours long and she was at the Bermuda airport to meet me with a big smile when I arrived. We left the airport on the eastern end of Bermuda in an autobus, known locally as a *jitney* (the only vehicles allowed on the island), crossed a causeway, and then drove west eight or nine miles to Hamilton Harbor. Hamilton is famous for its big fancy hotels, such as the Bermudiana, along the north shore of the harbor.

The taxi dropped us off at a ferryboat landing. Soon we were crossing the harbor towards the south shore less than one-half mile away. The ferry made two stops and then landed at Salt Kettle, Paget West, where we disembarked with our suitcases. After asking directions, we walked a few hundred yards to our small, rather secluded hotel called "Glencoe." Glencoe consisted of small bungalows surrounded by palm trees and lush vegetation with a path leading into a one-story main building with a small lobby and dining room. Outside was a large porch overlooking the harbor where we ate most of our meals. A path led down to a lovely sandy beach.

For nine days we had a wonderful time enjoying each other's company and seeing the sights of Bermuda. We rented motorbikes and drove east to St. George, the capital, where we toured King's Square with its famous stocks and pillory that were "designed to humiliate and humble erring citizens" back in the 1600s. We then motored on to Fort St. Catherine for a picnic lunch and more sightseeing. Our other fun activities on Bermuda included a round of golf at Riddell's Bay Golf Club, sailing a rented "Sunfish" around Hamilton Harbor, and a floor show at the Inverarie Hotel featuring Rudi Vallee. Rudi may have been a great singer and entertainer in the 1930s, but he was way past his prime in 1964 and terrible; his wife, 50 years his junior, was in Marcia's words, "absolutely dreadful and must be in the act so all her dresses could be deducted as business expenses." We also motorbiked around the island (Marcia got shocked one time when her leg touched the engine) and stopped to picnic and snorkel at a small secluded beach at the west end of the island.

At one point Marcia produced some brownies she had baked in New York to show me she could cook! Another day we were both amused by an item in the local newspaper about a court case involving the theft of "a pair of blue sneakers."

As the end of our vacation neared, we heard that Hurricane Dora was headed for Bermuda, but on September 6 and 7 it passed harmlessly some 350 miles to the south of us and then plowed into the east coast of the States. No sooner had Dora passed than we heard that Hurricane Ethel was headed our way. Every few hours I plotted Ethel on our hurricane-tracking map, and it appeared to be following in Dora's footsteps going west well to our south. Then, during the night, Ethel suddenly turned due north and headed right towards Bermuda. Everyone on Bermuda closed their shutters and boarded up their storefronts as the winds began to pick up. The next night, we were hit by hurricane-force winds as Ethel passed less than 100 miles west of Bermuda. The winds howled and the shutters banged all night. After the storm passed, we were glad to see that the only damage our hotel had suffered was fallen palm fronds all over the place.

The hurricanes delayed our flight to New York by a couple of days. Marcia sent a telegram to her roommate in New York telling her of the delay, but we found out later that it never reached her. So, nobody knew where we were during the hurricanes.

Just before we left Bermuda, I took Marcia down to the Glencoe beach on a moonlit evening and proposed to her that we get married. She said, "Yes!" Then we asked each other, "What do we do now?"

On September 13, Marcia and I flew to New York City. I checked into a hotel, and Marcia returned to her nice apartment located at Lexington and 72nd Street. Since I had last seen her in October 1963, Marcia had moved out of the Barbizon Hotel for Women into a fifth floor walk-up "apartment" in Greenwich Village, which shocked her parents, and then to her present fashionable east-side apartment with her roommate, Liz.

I called Mother and Jack in St. Louis, the Halseys in Tulsa, and my office in London to tell them I was going to be married. Since I had left London just to go on my annual vacation and nobody knew about my rendezvous with Marcia in Bermuda, the news came as quite a surprise to everyone. I asked my boss in London for extra leave time for the wedding to be held in Findlay, Ohio; he said I could work in the Findlay office to earn the extra time.

I spent about a week in New York City with Marcia and bought an engagement ring at Tiffany's that Marcia had chosen with a diamond weighing three-fourths of a carat. I came into the Marathon office the next day and said to Marcia, "Do you mind, I would rather you have a full carat diamond." So we took the first ring back and got another one. We also bought his-and-her wedding rings with inscriptions at Tiffany's. My gold wedding band was inscribed inside, "M.G.M. AND F.W.K. OCT. 17, 1964."

After a week in New York City, Marcia quit her job with Marathon Oil and flew to Chicago to meet her mother and father to look for a wedding dress at Saks Fifth Avenue. A wedding had just been canceled, so Marcia and her mother bought the wedding dress and all the bridesmaids dresses, which were just right for our wedding. There was the same number of bridesmaid's dresses as Marcia wanted, and most of them were the right sizes.

Marcia flew back to New York for a week. Her mother and father drove from Findlay to New York City, loaded up her stuff, and returned to Findlay. Meanwhile, I went to St. Louis and Tulsa to tell everyone about the wedding plans and then flew back to Findlay, where I worked in the Marathon office for a week to extend my home leave.

Marcia and I went to the courthouse in Findlay and obtained a marriage license, and the date of the wedding was set for October 17, only some three weeks away. Leading up to the wedding, there were several luncheons and a bridal shower for Marcia. Our good friend, Jerry Overall, hosted a party for both of us, and Everett Dill threw a bachelor party for me. Marcia's mother obtained a copy of the wedding vows from the First Presbyterian Church, my church, and made several changes to make it compatible with the Marcia's Christian Science Church. I took my shoes to a shoe repair shop and had the soles dyed black so they would look dark when I knelt at the altar, as advised in our wedding book.

Uncle Marian, Aunt Peg, Mother, and Jack all drove up together from St. Louis to Findlay. Marcia's aunts and uncles came in from Chicago. On Friday, October 16, we had a wedding rehearsal at the First Presbyterian Church. Mother and Aunt Peg rushed up and hugged Marcia, but Marcia was confused because she did not know which one was my mother. Following the rehearsal was a rehearsal dinner for some 30 people at the Findlay Country Club given by me, as bridegroom, and my mother. Actually Uncle Marian made most of arrangements and paid for the affair too. At the dinner Uncle Marian gave a toast to Marcia and me and then gave a long, glowing tribute to my mother.

On Saturday morning, October 17, the day of the big wedding, I went to a local barbershop where I got a haircut and a very close shave for the occasion. Meanwhile I was memorizing my part of the wedding ceremony so I would not make a mistake. Then, I put on my rented tuxedo with tails (what I have always called a "clawhammer coat" from something I read somewhere). And then, the wedding took place. It can best be described by quoting from the lengthy article that appeared in the Findlay *Republican Courier* newspaper the next day:

> *Touches of autumn beauty marked the Saturday marriage of Miss Marcia G. Mehl to Frederick W. Kelly, Jr., London, England. The Rev. Frederick Allen performed the double ring ceremony at the First Presbyterian Church.*
>
> *The bride is the daughter of Mr. and Mrs. Alfred H. Mehl, 801 Fifth Street. The bridegroom is the son of Mrs. F.W. Kelly, Sr., St. Louis, Mo., and the late Mr. Kelly.*
>
> *Preceding the 3:30 p.m. service, music was offered by Mrs. Earl Potts, organist, and Mrs. Wilbur Hall Jr., soloist. Mr. Mehl gave his daughter in marriage before an altar graced with white gladioli, white mums, palms, and candelabra. White bows marked the pews.*
>
> *The bride chose a gown of light ivory peau de soie and reembroidered Alencon lace. She carried a white orchid surrounded by white pompons and Fiji mums with touches of white pearlized grapes.*
>
> *Miss Nancy Eversole, Chicago, Ill., served her cousin as maid of honor. Bridesmaids were Miss Rosemarie Wanko, New York City; Miss Dianne Cummins, Cincinnati; and Mrs. William Hillshafer Jr., Washington, D.C. The attendants wore identical ankle-length gowns of antique gold silk crepe, and each carried a cascade of bronze mums with gold and rust daisy pompons.*
>
> *The best man for the bridegroom was the brother of the groom, John Kelly, St. Louis, Missouri. Seating the 150 guests were Everett Dill, Presley DeJarnett, and Jerry Overall, all of Findlay.*
>
> *Following the wedding, Mrs. Mehl headed the receiving line followed by the bridegroom's mother. Each had a corsage of two green cymbidium orchids.*
>
> *A reception was held in Fellowship Hall at the First Presbyterian Church. Mrs. Wayne Mizsak was at the guest book. The buffet table was decorated with gold and rust mums. The center for the*

decorations was a four-tiered wedding cake with white, gold, rust, and yellow pompons between the tiers.

For her going-away ensemble the new Mrs. Kelly chose a three-piece coral wool outfit with a black velvet collar.

She is a graduate of Findlay High School and attended Northwestern University in Evanston, Ill., where she was affiliated with Kappa Delta Sorority. She also attended Findlay College and was employed as a secretary at Marathon International Oil in New York.

The couple will be at home after October 19 at 55 Park Lane, London, England.

Marcia and I like to recall with a chuckle an elderly man that came through the receiving line after the wedding. He had a stopwatch in hand and said, "Not bad, not bad. Only ten minutes and thirty seconds!"

THE HAPPY BRIDE AND GROOM.

THE BRIDE, GROOM AND PARENTS.
From the left: Marcia's father, Alfred Hugo Mehl; Marcia's mother, Grace
Anton Mehl; the bride and groom; and my mother, Dana Milleson Kelly,

Following the wedding, Marcia and I went back to her mother and father's house at 801 Fifth Street for tea with family and close friends from out of town. At that time, we opened our wedding presents in the garage that served as a kind of screened-in porch. Then I went back to my hotel. Marcia finished packing for the trip to London surrounded by her friends in her tiny bedroom as they all laughed and joked. Marcia had so much in her suitcase that she couldn't get it closed; her father had to come back later and repack her suitcase for her.

At about 9:00 p.m. that night I picked up Marcia at her house in my company car and we drove north to Toledo, Ohio. An hour later we checked into the Continental Inn in Toledo, which was not very glamorous but conveniently located near the airport. At 5:00 p.m. the next day, we flew from Toledo to New York, where we made a connection for our flight to London. We arrived in London early on Monday, October 19, and then spent our first day together in my service flat at 55 Park Lane.

The Marathon Oil Company office in London had a wedding shower for us and gave us some Waterford crystal cocktail glasses. Soon thereafter, all our wedding presents arrived from Findlay. Marcia's mother and father had had to pack them up for us and ship them. As Marcia recalls, "They had all the hard work to do." We rearranged the flat with our wedding gifts and Marcia added other nice, homey touches. The African and Libyan momentos came down off the walls. Marcia started learning how to cook in a tiny kitchen only some six feet square. We dined at nice restaurants in the West End of London, had tea at the famous Dorchester Hotel next door, and had coffee and newspapers on Sunday morning at the Hotel Intercontinental just down the street.

Marcia became a bit starved for sunlight in our flat because it was continuously in the shadow of a large building next door. The only way she could see daylight was to lean out the window and look towards Hyde Park. And, just outside our windows there was a historic plaque attached to the wall that recorded that Florence Nightingale (1820-1910), famous as the nurse that reformed hospitals during the Crimean War, once had a hospital on this spot. Occasionally, an itinerant Irish tenor would sing in the street under our windows and we would throw him a few coins.

Our building at 55 Park Lane was such a prestigious place that many prim and proper wedding receptions were held in the front lobby. I had to put on a coat and tie every time I left the building as I often found myself mingling with guests in full dinner jackets and evening gowns. I even had to put on a tie to empty the garbage down a building corridor because all the porters wore formal attire with neckties and I might run into one of them in the hall. On one occasion, Marcia and I noticed water leaking from the ceiling in our bathroom. We both got dressed up and went upstairs and knocked on the door of the flat right above ours. An elderly British couple came to the door and was completely flabbergasted to find a strange American couple on their doorstep unannounced! I explained the situation and for a moment they were completely incredulous that we were there talking to them. Finally, the wife went to their bathroom and came back to report that water had indeed overflowed onto the floor. They gave us an embarrassed apology and thanked us for telling them.

Marcia and I got settled into our happy, newly-wed life. And, we really got to know each other, after our Trans-Atlantic letter romance, as we exchanged stories about ourselves. Here is Marcia's story about herself:

My father, Alfred Hugo Mehl, married my mother, Grace Louise Anton, in Chicago on February 16, 1929. I was born in St. Joseph, Missouri, on December 23, 1938. Mother and Dad were unable to have children of their own and "found" me at the Sheltering Arms with Mother Zimmendorf in St. Joseph, Missouri, when I was six months old. I spent the next 10 years in the Chicago area: Edison Park (Oconto Ave.) and Park Ridge (corner of Touhy Avenue and Knight). My Aunt Ruth and Uncle Frank Nance lived across the street from us in Park Ridge with my cousins Joanne and Bob. During those approximate seven years in Park Ridge, my aunts, Carol Meyer (and cousin Keith about 6 months old), Janet and Eloise, lived with one of us, or both, at some time or another. Mother's family was extremely close and, with her being the eldest, most holidays were spent at our home with all the family gathered around.

Dad worked for the Chicago and Eastern Illinois Railroad (C&EI) and during World War II was a Traveling Passenger Agent on their Dixie Flagler streamline train, which ran between Chicago and Jacksonville, Florida. Occasionally Mother and I would go with him on his "route" and spend several days in Florida until he returned to get us. Usually he was "on" three days and "off" three days. Dad worked for the railroad for 25 years. I loved riding on the train and still do to this day. In 1947, he joined Chicago Desk Pad Company as a manufacturer's representative selling office supplies under the "C-Line Brand" in Missouri, Oklahoma and Arkansas.

In 1951, when I was twelve, my parents and I moved to Findlay, Ohio, where Dad's C-Line territory covered Ohio, Michigan, Indiana, Kentucky and West Virginia. Mother and I occasionally accompanied him on his business trips during summer vacation. I especially enjoyed going to Michigan where we would plan on being in Traverse City for the Cherry Festival.

I spent the remainder of my formative years in Findlay, attending junior high and high school there.

Being a small town at that time (approximately 16,000 residents), Findlay was a wonderful place for a family to live and this enabled me to be editor of the school newspaper, cheerleader *(both in junior and senior high school), and part of the Homecoming Court. Since my circle of friends was extremely social, we joined every club in school and were constantly planning parties and having a wonderful time. From the beginning of junior high we always had coke parties before every football game and continued that through high school. Great fun! When we first moved to Findlay our home on Stadium Drive was only a block away from Donnell football stadium (and also Donnell Junior High --named after the Donnells, who ran Marathon Oil). Our home was the logical meeting place to gather, which we did often.*

After high school, I attended Northwestern University planning on a journalism career leaning towards the advertising aspect. And, I joined the Kappa Delta Sorority. However, after a short time, and after visiting a large advertising agency in Chicago, I discovered that advertising was mostly a male-dominated occupation at that time (in the 1950's). I then decided to head back to Findlay to attend the local college and entered their business school.

While attending Findlay College part-time, I worked as a bookkeeper at Eoff Insurance Company, whose owner was a friend of the family. I don't think I was terribly good and really didn't enjoy it very much.

I worked for Eoff Insurance for about nine months and then had an opportunity to go on a Findlay College tour to Europe. Wow, did I jump at that! I would even get college credit for having fun, so that was really my kind of class! It was truly a dream come true. This may have been the beginning of wanderlust.

After a wonderful month in Europe, I returned to Findlay College in the fall. While attending classes, I worked part-time in the college business office, which was quite enjoyable.

Also at Findlay College, I joined the choir and during Easter vacation we toured in Pennsylvania. Since our college was a Church of God college, we only sang in Church of God churches on our tour. We stayed in the homes of members of the various congregations and had a wonderful time.

Findlay was the home of The Ohio Oil Company (now Marathon Oil) and at that time it was the main industry in this small Midwestern town. Most of my friend's families had some affiliation with "The Company," as it was called among the locals. Fighting the trend, I tried not to get involved with "The Company," but discovered that that was where the opportunities really existed.

I worked for Marathon in Findlay for about three years and then, as a single person, got restless and looked elsewhere for more adventure. As it turned out, Marathon Oil Company was expanding their New York office and I investigated the possibility of transferring there.

At that time, it was unheard of for any female to move from the home office to any other area unless it was with a spouse who had been transferred to the new office. As a result, I had to quit my job in Findlay and was re-hired in New York. Fortunately I was able to keep my benefits so it wasn't a big deal to me but it was to Marathon.

I felt I could be a big asset to the New York office because I knew many of the executives from Findlay and I could add an "at-home" feeling. I loved it from the very beginning. I lived at the Barbizon Hotel for Women for about nine months and then sublet a fifth story walk-up in Greenwich Village from a friend who was going to Europe for three months. What fun!

After my friend's return, I then moved in with another girl who was looking for a roommate. I think our apartment was located around 73rd and Lexington. And, that's where I was living when we got married.

Life was very good for Marcia and me at 55 Park Lane in London. Then, one day I reluctantly took Marcia aside. I explained to her that, as an international oil company geologist, our life at 55 Park Lane was at the very top of the scale and was probably the best situation that could ever be expected. I cautioned her that any of my future assignments would probably not be so nice. Little did I know at that time that our next post would be at the very bottom of the scale in Karachi, Pakistan!

THE NEWLY WEDS AT HOME IN LONDON.
Marcia and I at home in our service flat at 55 Park Lane, London,
next door to the Dorchester Hotel. Our building was so formal
that I had to wear a coat and tie to empty the garbage!

22 ✳

Assignment Ankara and
a Baby Expected

In November 1964, my office in London got the word that the Turkish National Oil Company, known as "TPAO," had just released all the geological data that it had collected over many years from international oil companies that had previously explored for oil in Turkey. These files had been confidential, but were now openly available in Ankara to all companies interested in new exploration ventures in Turkey. I was put in charge of making a trip to Turkey's capital city, Ankara, to bring copies of all the data back to London. Needless to say, Marcia and I were distressed that I had to leave on a long assignment so soon after our wedding. But, I explained again that this was the life of an oil geologist.

A few days later I was on my way to Ankara, accompanied by my British colleague, Rodney Collomb. After changing planes in Rome, Italy, we finally landed at Ankara. The airport was on the top of a snow-covered plateau, and we took a taxi down into the city of Ankara, located in a foggy and smoke-filled valley. We checked into a modest but nice, clean hotel and bought city maps so we could find our way to TPAO's offices. The next morning we walked to the TPAO office and met with the managing director for exploration. He was very cordial and spoke excellent English. He said we were free to examine and make copies of all the newly released geological, geophysical, and drilling data. We were astonished when he told us the large amount of data involved! He then showed us a small room filled with gray steel filing cabinets containing the data. The director told us that we could make copies of all the data, but he gave us a shock when he said that there were no copying machines in the building and we could not take any data out of the building.

After leaving the TPAO building, Rodney and I visited several office supply stores to see if they had any portable copying machines we could rent. We were dismayed to learn that they not only did not have any portable machines but that there were very few copying machines of any type in Ankara.

Back at our hotel, Rodney and I pondered our situation and decided to try to photograph the data. The next day we hired a professional photographer and he brought several top-of-the-line 35-mm cameras to the TPAO office. The photographer set up his best camera on a small tripod pointing down at a table and arranged several floodlights under the camera. We took some geological reports out of a filing cabinet, unbound them, and started placing pages under the camera as the photographer snapped the shutter. At the end of the day we had photographed a hundred or so pages and the photographer took the film back to his studio to develop and print it. The next morning we stopped by to pick up the photographer to resume work, but he looked very distressed. On developing his film, he had found that about every tenth picture was blank. He concluded that the shutter mechanism on even the best 35-mm camera is not designed for fast, continuous exposures and that the spring must "rest" between shots. Sadly, we decided that 35-mm cameras could not reliably copy all the thousands of pages of data.

Rodney and I brooded over our ever worsening position as failure of our mission loomed on the horizon. Then I got an idea. I recalled that the 8-mm movie camera I had used in Libya had a single exposure setting on the shutter. I wondered if we could find a photographer with a 35-mm movie camera that had single-frame shutter capability. The next day we went to the TPAO director and explained our predicament. We asked him if he could locate a movie photographer who could help us make our copies. I told him that if we could photograph the data on 35-mm movie film, we would pay for a set of copies of the reels for him to keep. At our next meeting, the director announced that he had found a Turkish government movie producer who might be able to do the job for us. We took a taxi out to a group of drab government buildings and had coffee with the moviemaker. He told us he made documentary films for the Turkish government and did have a portable 35-mm movie camera with a single-frame shutter setting. He said he could take two weeks of vacation to work with us if his camera would do the job. We then agreed on the fees that we would pay him.

The next day we met the movie producer at the TPAO office, and he set up his movie camera on a tripod with lights. Then we started photographing pages of data one at a time until we had several hundred test pages done. The producer took the film back to his laboratory to develop it. The next morning the producer was all smiles as he told us the pages had photographed perfectly and with no blanks. We were now ready to get to work. The three of us settled into a daily routine of unbinding reports and placing the pages one at a time under the movie camera.

In the evenings Rodney and I went shopping for leather goods and handicrafts, which were of excellent quality and inexpensive. We did some sightseeing, but other than a statue of modern Turkey's founding father, Kemal Ataturk, the city was mostly dirty yellow, square, concrete buildings from the 1920s and 1930s (what I jokingly called in Libya "early Mussolini period" architecture). We tried various restaurants in Ankara and found the food different but not very outstanding. The weather continued to be cold and smoggy. We were told the thick smog was caused by the low-grade soft coal burned as fuel throughout Ankara. One evening we went to a restaurant and nightclub that advertised "belly dancing" as the floorshow. The place turned out to be a huge smoked-filled club packed with men sitting around tables in their overcoats. Turkish music played as all the men sat silently sipping coffee waiting for the floorshow. We surveyed the glum crowd and the far distance to the dance floor and decided just to have a drink and go back to our hotel.

Finally after almost two weeks we finished photographing the typewritten pages of data. We then turned our attention to the maps of all different sizes. We attached a large white backboard to a wall and focused the movie camera and lights on it. One by one we unrolled the maps and pinned them to the wall. Some were so large we had to photograph them in several pieces.

Meanwhile I was talking almost daily to my new bride in London, and she was not very happy about my being away so soon after our wedding. This was her first introduction to the duties of a wife of an international petroleum geologist. We had been married only three weeks before I left for Turkey, and we had not yet set up bank accounts in both of our names or made any other financial arrangements. As a result, Marcia had to get frequent cash advances from my boss in the London office, especially as Christmas was approaching and overseas packages had to be mailed several months early to ensure their arrival in the U.S. in time for the holidays.

Finally Rodney and I finished all the maps, and it remained only to duplicate the reels of data for TPAO, pay the photographer, and thank the director of TPAO for his hospitality. At this point I turned everything over to Rodney and headed back to London. After a thorough body search by the police at Ankara Airport, I boarded my BOAC aircraft and flew to Rome, Italy. I had a long layover in Rome and then flew across Europe towards London. After a long, boring flight, the pilot said we were approaching Heathrow Airport at London but that the fog was getting very thick. Through small openings in the fog I could see the welcome lights of London, but the pilot suddenly announced that the fog was too thick to land and we were diverting to Manchester, England, some 175 miles away to the northwest.

Our plane landed in Manchester late on a cold, wintry night. BOAC then informed us that British Rail was "putting together cars from sidings" to make up a train to take us to London. An hour or two later we were bussed to the waiting train and took our seats for London. The train was practically empty, dark, and very cold. We all kept our hats and coats on to keep warm as the train crept towards London. I was tired, cold, and feeling sorry for myself when I suddenly remembered I had a bottle of duty-free brandy in my briefcase. That brandy warmed my body and my spirits for the long trip to London, where my new bride eagerly awaited my return.

but no air-conditioning as was usual in London. We had an entrance hall, a living room that led through a small arch into the dining room, a kitchen, utility room, two small bedrooms, a master bedroom, a nice large master bathroom complete with a European style "bidet," or sit-bath, and another bathroom. A couple of very friendly porters took care of things around the building.

We had all my old bachelor furniture shipped over from Findlay, Ohio, where it had been stored. We even ended up with my large oil painting of Libyan camels and date-palm trees hanging in the living room, much to Marcia's chagrin. At least this would do until we bought some proper furniture and furnishings, which would be fun to do in London. We had a little balcony outside our picture window where we would sit and watch the activities outside. Bayswater Road was a busy and interesting place. There were frequent double-decker London buses going by and the upper decks looked right into our flat. Every Sunday, artists hung their paintings on the park's iron fence across the street as hundreds of people walked by. Large public demonstrations paraded by for various political causes. And, one time when a friend of Marcia's left our building she was accosted by an irate woman who told her in no uncertain terms that this block was her territory!

As September passed, we looked forward to the arrival of the baby. Marcia took "natural childbirth" classes under the instruction of a Ms. Erna Lowe, who had published a book on her methods, and I participated in the last class myself. In the meantime, we furnished one bedroom as a nursery, and Marcia bought a beautiful, pre-owned, mahogany, baby-grand piano from Harrod's Department Store that had been made by The Challen Company in 1926. And so, in the late summer of 1965, life was very good for the Kellys of Bayswater Road.

And now, on the wise advice of my editor, Peggy Stautberg, I must conclude this book or risk never getting it finished. However, the adventures in my life certainly did not stop at this point and I plan to write "the rest of the story" in the near future. Just so you will look forward to the next volume of my autobiography, I am going to give you a short preview of coming attractions in the next and final chapter.

23 ✳

Preview of the Rest
of the Story

My worldly adventures certainly did not stop with my move to London and marriage to my dear wife, Marcia. In my next book, if I have the time and energy to write it, I will have many more strange and amusing stories of foreign lands to tell. However, these stories will take a new direction because they not only involve Marcia and me, but also our three dear children, as they are added to our family. Just to whet your appetite, here is a brief preview of coming attractions.

My book on our years in London (1965-1973) will start with the birth of our son, Rick (Frederick William III). Stories about our first years in London will include our trip up the Loire River of France, picnics in Hyde Park across the street from our flat, my night school classes at The Polytechnic of London for a Diploma in Management Studies, and my continuing evaluation of the oil potential of Italy, including field trips there.

Our first daughter, Heather Anne, is born. And, around this time, a brutal double murder happens in our apartment building. My oil evaluations widen to involve the West Coast of Africa.

In 1970, my work starts to focus on Pakistan, a Moslem country that was previously part of British India. For three years, I make frequent trips from London to Rawalpindi and Islamabad, Pakistan, with my boss and an attorney to negotiate a petroleum concession with the Pakistani Government. You will read about all of our frustrations as well as the large quantities of duty-free Glenfiddich Scotch whisky we consume on these trips.

Meanwhile, back in London, Marcia finds a charming, fully furnished cottage on a farm on the South Downs in West Sussex that we rent for £35 a month. We buy a really neat, second-hand London taxicab to use for driving down to the cottage. Marcia becomes the first woman to drive a taxi in London (although not for hire!) and cabbies stop to get her autograph. The taxi is also very handy in London, as we are able to take friends to the theater and just park it on the sidewalk like an out-of-service taxi.

MARCIA DRIVING OUR TAXI AT OUR COTTAGE IN SUSSEX.
Marcia at the wheel of our Austin taxicab in front of our
country cottage in West Sussex. She was the first woman to
drive a taxi in London (although not for hire, of course), which
gained her much notoriety amongst the taxi drivers.

At The Polytechnic, I learn how to program an IBM computer in the FORTRAN computer language using punched cards and get hooked on computers for the rest of my life.

In addition to Pakistan negotiations, I study the oil potential of Greenland and Spitzbergen and make nice trips to Copenhagen, Denmark, and Oslo, Norway. I also spend some time as a wellsite geologist on a rusty, old drill ship (it was formerly an iron ore carrier on the Amazon River, and I called it "The African Queen") that was anchored off the southern Irish coast near Cork. This well discovers natural gas that is piped into Ireland.

Rick and Heather attend a very "English" private school called Connaught House near Marble Arch. At home, we have our "daily" housekeeper, Nellie, who is also an excellent baby-sitter and practically becomes one of the family. Our social life is very pleasant with such events as a charity champagne dinner at The Tower with a private viewing of the British Crown Jewels, and tea with the Headmaster at Eton College.

In 1971, I sell an exploration deal to Shell Oil Company to drill a wildcat well in Northern Ireland, and this starts a new round of adventures for me, including having pajamas hanging in three different places. For this operation, Marathon appoints me Vice President and Resident Manager for Northern Ireland. Our field headquarters are in an old castle converted to a hotel located northeast of Belfast, and my colleagues and I work our way through the hotel's wine cellar and drink numerous Irish Coffees every evening. We abandon the well as a dry hole just as the Irish Republican Army (IRA) starts blowing up Belfast, including a car bomb that blasts all the windows out of my Belfast office.

As my negotiating trips to Pakistan continue, Marcia, the kids and I enjoy life in London and our weekend trips to our cottage in Sussex. The kids and I find pieces of Roman pottery on a hill near the cottage. Rick and Heather ride horses up on the chalk hills. Marcia hunts for antiques. I make a trip to a small village near Southampton where I buy a small oil painting of the RMS *Titanic* with a luncheon menu from the ship pasted on the back that is dated April 14, 1912, the day the ill-fated ship hit an iceberg and sank. I pay the equivalent of $40 for it and the owner certifies that he bought it from the son of one of the surviving *Titanic* crewmembers. [Little did I know then that in 1989 and 1999 I would appear on television with this menu on the *Antiques Roadshow* and the *Oprah Winfrey Show*, and it would later be sold at auction for $65,000.]

On April 9, 1973, I travel with my team to Islamabad, Pakistan, where we finally sign our concession agreement to explore for oil in a 10,000-square-mile area straddling the Makran Coast of Pakistan, adjoining Iran. Marathon names me as Vice President and Resident Manager of Marathon Petroleum Pakistan Ltd. to run the operation. I make the first visit to our concession area by Pakistani commercial airliner. What is supposed to be a day-long round trip turns into another adventure when our plane gets caught in a sandstorm. We are forced to spend the night in an ancient mud brick fortress in the village of Gwadar. I eat goat with my hands and sleep in Pakistani-style pajamas on the roof along with my fellow Pakistani passengers.

Marathon assigns me to move with my family to Karachi, Pakistan, to set up and manage our oil exploration project on the Makran Coast. In other words, through my negotiating efforts, I manage to move our family away from our pleasant life in London to a strange, Third World culture and lifestyle in Pakistan. As I often say later, "Some people will do anything to become resident manager of their own operation."

In a future book, which I intend to call "The Karachi Chronicles," you will read all about the Kelly family's six years in Karachi. Our little family arrives in a hot, humid and smelly Karachi on July 5, 1973, with no house, no office and no local staff to help us. The "culture shock" is huge, especially for Marcia, as women are not treated very well in Moslem countries. From the start it seems like we have a crisis of some kind at home or at the office almost every day for six years, but we have some fun too.

We move into a nice bungalow covered with bougainvillea flowers located at 127 KDA Scheme No. 1, Karachi. Our furniture arrives by ship and we cover the bare terrazzo floors with colorful Pakistani carpets. Rick and Heather enjoy the Karachi American School, where they have good friends (from America, Pakistan and around the world), good weather for sports almost every day, and little homework. For family recreation, there is the Yacht Club, tennis, golf (cheapest in the world at 35 cents per round; but with oiled-sand greens and hard-pan fairways that is all the course is worth), a broken-down beach hut on the Indian Ocean, crab fishing in Karachi Harbor, and movies on our rooftops provided by the American Consulate. Marcia plays bridge and Mahjong. We make some wonderful American, Pakistani and European friends and attend frequent dinner parties with much fun and abundant alcohol.

You will learn how Marcia is kept busy and frustrated managing a houseful of male servants. Since Pakistan was formerly part of British India, most of the servants speak just enough English to get into trouble. Or, as Marcia says, "Happiness is a servant that only asks twice a year to visit his village because his mother has died." Our "sweepers" clean the house, our "houseboy" polishes the brass, our *mali* tends the gardens, and our *chowkidar* guards our gate every night. A tailor comes by occasionally to sit cross-legged on the porch and make our hot weather clothes on a hand-powered sewing machine (there are no imported clothes available in Pakistan). The *dhobi* does our laundry.

All our food is bought at the Empress Market, built by the British in 1889, which is a large, sprawling and dirty complex of tiny shops selling spices, fruits, vegetables, canned goods, live chickens, and meat. And, everything is covered with flies. Marcia always goes to the market herself, and upon arrival hires a "coolie" to fetch things for her, especially from the smelly meat market. The meat consists of ragged cuts of lamb, beef and water buffalo hanging on hooks and covered with flies. There is no refrigeration so the meat is butchered the same day that it is sold. Men chop-open animal skulls for the meat inside. The only part of the scrawny cows that we can eat is the cylindrical tenderloin and, therefore, all our beef for six years is served as round slices of tenderloin, but some not so tender. There is also a fish market where we can buy fresh fish and very large, striped, "tiger" shrimp. My boss from Findlay, Ohio, on touring the Empress Market shakes his head and says to me, "Fred, this is where they separate the men from the boys."

**EXPLAINING OUR EXPLORATION PROGRAM TO THE
PAKISTANI MINISTER AND SECRETARY FOR ENERGY IN 1975.
As Resident Manager, I spent much of my time working
with officials of the Government of Pakistan.**

Our "bearer" serves every meal to us at table. Our cook, who has worked for the British and has a very good disposition, boils and filters all of the drinking water, and soaks all of our vegetables in "pinky solution" to sterilize them. There are no imported foods or mixes in Pakistan, so our cook bakes our bread and pastries from scratch every day. You will read about the time Marcia orders twelve "Butterball" turkeys raised at a local U.S. government farm, but she is shocked when they are delivered alive and run all around our backyard (until our cook slaughters them on the spot). In spite of the obstacles, we have good meals every day, thanks to Marcia's careful management.

We also enhance our diet of local produce with occasional items that I buy on the "black market." I rather enjoy this, as it is a kind of daily treasure hunt for me. I go to Empress Market and ask the shopkeepers what black market, foreign foods they have for sale. They produce things like caviar, frozen (how many times?) pork chops, hot dogs and canned hams, all of which have been stolen from airlines, foreign consulates, or other foreigner's homes. One time I am offered American breakfast cereal that I am assured is very fresh because it has just been stolen from the American Consulate that morning.

Pakistan, being a Moslem country, restricts the sale of alcohol to foreigners in small amounts. As resident manager, I need large quantities of wine and liquor for parties and so I buy lots of bootleg liquor. One of my regular smugglers drives up to my office during working hours in a truck and unloads several egg crates full of straw into my conference room; under the straw are bottles of wine and booze. One night I drive to a Pakistani's house and in the dark buy ten cases of illicit American beer for cash (the Pakistani could be flogged in public if caught doing this).

You will hear how our family flirts for six years with malaria, amoebic dysentery, hepatitis, tuberculosis and rabies, but never contracts any of them, thanks to pills and shots, Marcia's careful attention—and a lot of luck.

Our American comptroller, Jerry Zimpfer, and I rent a bungalow and set up an office. Our American staff organizes field exploration operations. We do surface geological surveys (using rented World War II radios and Morse Code to communicate) and then seismic surveys. We hire local staff and are soon up to 21 people. In two years, we are ready to drill our first well. We bring in some modern short-wave radios but they are confiscated by a radio inspector due to improper licenses, until I threaten to bring the government in Islamabad down on him.

Marathon drilling staff arrives from the States with their families and our office expands to 35 people. A drilling rig is shipped in from Dubai. And, because there is no harbor near the drill site, we use World War II landing ships, which hit the beach at full speed and then open bow doors for the rig to be dragged out on to the sand. It is sort of like the Marines hitting the beach at Iwo Jima in World War II, only we also have poisonous sea snakes in the water. The well is a total dry hole with no indications of oil or gas, which is discouraging.

Meanwhile, our family is doing fairly well in Karachi in spite of the many hardships. We have dinner parties at the Sind Club (founded by the British in 1871; no women allowed in the bar) and good meals at the Boat Club (founded by the British in 1881). We throw large parties at home, and attend many, many American community affairs on the grounds of the Karachi American School (I have always said it was like living in a "commune"). Every year we have home leave back to the States and visit most of the countries in Europe on the way. The first thing we always do on arrival in Europe is buy large quantities of foods we cannot get in Karachi, such as salami, sausages and cheese, and have a feast in our hotel room.

Then, in the late summer of 1976, Marcia and I get the exciting news that another addition to the family is expected. As we plan for the birth, we ask Marathon if Marcia can go to London to have the baby. Marathon gives approval for this but with two other children to look after and with the delivery date uncertain, Marcia wonders if going to London is practical.

Following our onshore dry hole, we shift our oil exploration efforts to the offshore areas of the Makran Coast. We add another 10,000 square miles of concession area and carry out offshore seismic surveys. In December 1976, we float-in an offshore jack-up drilling rig. This type of rig jacks itself up off the water on four legs. We start drilling. However, the well encounters high-pressure and is reluctantly abandoned. This turn of events now puts our entire Pakistan venture into question. And, on top of this, most of our American personnel, including Marcia and me, are so fed up with life in Pakistan that we are ready to get the heck out of there anyway.

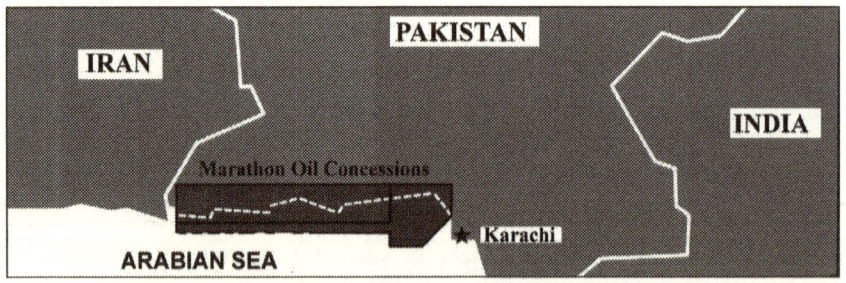

MAP OF OUR PETROLEUM CONCESSIONS IN PAKISTAN.
Shown on this index map are Marathon Oil Company's two
concession areas that we explored for oil from 1973 to 1979 without
success, and the city of Karachi, our home for six long years.

Meanwhile, at the office we do more marine seismic work and try to decide whether or not to drill another offshore well. Then, in February 1979, campaigning and demonstrations start for the March national elections to elect a new parliament and Prime Minister. The present Prime Minister, Zulfikar Ali Bhutto, is very powerful and appears certain to be re-elected, but then, to everyone's surprise, all the opposition parties band together to oust him. Rioting between parties starts in the streets of Karachi as crowds fight each other and burn buses and government buildings. The military finally declares a curfew and shuts the city down. We have "curfew parties" at the American School.

In late February, Marcia begins having labor pains and we make a couple of false runs to the Seventh Day Adventist Hospital. On one trip, a crowd of rioters stops our car and threatens us with rocks, but our driver explains who we are and we make a U-turn and get back home safely. On March 5, the election rioting stops as the parties await the results of the March 6 election. During this brief lull, Marcia suddenly knows the baby is about to arrive and we safely drive to the hospital. While Marcia is in labor at about midnight, I start getting calls at the hospital from our friends asking me if the baby has arrived yet. I keep reporting that nothing has happened. Then, at 5:12 a.m. on March 6, Christine Elyse is born. We find out later that the reason I have gotten so many calls was not out of concern about us but because our friends were having a party and had bet on what time the baby would arrive. We hire Mary, a very nice, local *ayah*, or governess, to look after Christine.

Following the election on March 6, the rioting begins again and gets very violent because the opposition parties find out that Bhutto has won and they accuse the authorities of rigging the election (I jokingly say that Bhutto won with 110 percent of the votes.) Our only news about the rioting in our own city comes from listening to BBC short-wave broadcasts from London as the situation gets worse and worse.

At this time, the Americans on my staff and their families become increasingly worried about their safety and it is up to me to decide whether or not we need to evacuate the country. The U.S. Consulate advises me to stay put for the time being. Finally, I call Bill Swales, President of Marathon Oil in Findlay, and tell him we are on the verge of evacuating. His advice to me is, "OK, but don't wait too long!" The next day, July 5, the military declares martial law, takes over the country and declares a curfew allowing only a couple of hours of shopping per day. General Zia-ul-Haq becomes the head of state. Former Prime Minister Bhutto is sentenced to death and we are greatly relieved when no rioting breaks out when he is hung in April 1979.

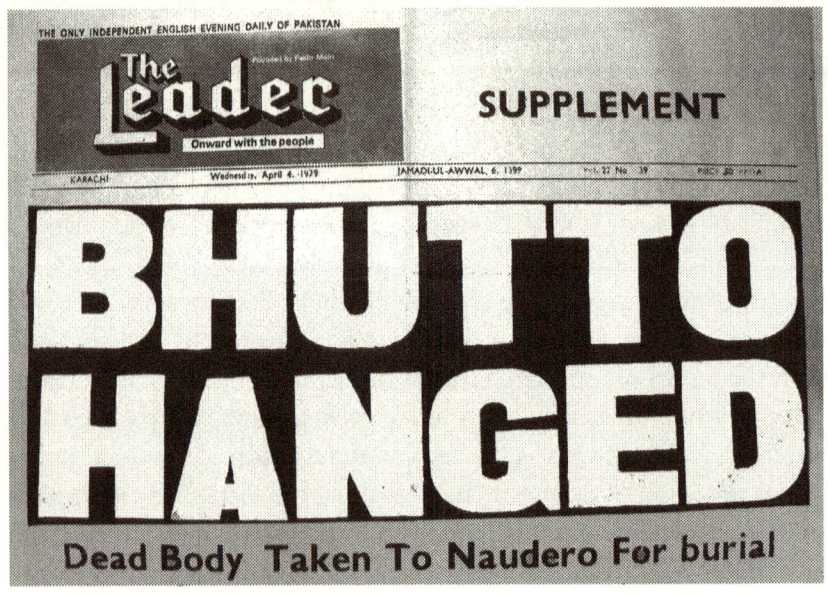

NEWSPAPER HEADLINE PROCLAIMING THAT FORMER
PAKISTANI PRIME MINISTER BHUTTO HAS BEEN EXECUTED.
**After being arrested by the new military government, Zulfikar
Ali Bhutto was found guilty of murder and hung on April 4,
1979. This followed a month of rioting and curfews and we were
greatly relieved that the country became quiet after this event.**

In view of our totally disappointing drilling results and the worsening political and security situation, Marathon decides to shut down the Pakistan operation. I have the unpleasant job of terminating our local employees, some of whom have been with us for over five years, and closing the office. We have spent millions of dollars on our venture without any luck, but that is how the oil business goes sometimes. I am proud, however, that I have managed a very tough operation for six years without any real disasters, and I did it in a remote part of the world with very little supervision from our home office in Findlay, Ohio.

In June 1979, Marcia and I make our final departure from Karachi with our three children. After several weeks of vacation, it was time for me to fly to Houston, Texas, to take up my new assignment there as an international contract negotiator. Marcia arrives in Houston a week later and we buy a house in west Houston, where we still reside 25 years later.

The details of our years in Houston will be the subject of yet another future volume. To summarize my career during this time: I negotiate contracts and scout out contract opportunities in places all over the world. I sign an exploration agreement in Luanda, Angola, on behalf of Marathon. Then in 1984 Marathon reorganizes the international exploration division and puts me in charge of international governmental affairs. I evaluate the political and security risks for every new exploration venture in which Marathon considers investing, including China, and also keep an eye on the countries in which Marathon already has oil production. I really love this job and save the company lots of money by warning them about major political or security problems in countries in which they had intended to explore. Then, in June 1992 I am offered a retirement package that I cannot turn down. And so, I retire after 38 years of service with Marathon Oil Company, which extends all the way from my college graduation to my retirement.

And, so it is now time to close this memoir and say farewell—or, as they say in Libya, *ma salaama*, or in Pakistan, *koda hafez*. Thank you for staying with this book all the way to its conclusion. I hope you have enjoyed your trip through time and space with me from my origins and ancestry up to the present day.

**RIDING A CAMEL AT THE GREAT WALL OF CHINA
SHORTLY BEFORE MY RETIREMENT IN 1992.**

About The Author

Fred Kelly was born in St. Louis, Missouri in 1931. He graduated from The University of Tulsa with an engineering degree in geology in 1954. From there, he embarked on a 38-year career in international oil exploration with Marathon Oil Company, including 22 years overseas in Libya, England and Pakistan. He has traveled all over the world both on business and for personal adventure to places including the Valley of the Kings in Egypt, Mexico's "Well of Death," and the Ubangi tribal area of Central Africa. Fred's other interests include computers, amateur radio, scuba diving, archaeology, writing and genealogy. Fred and his wife, Marcia, still travel frequently—especially when it comes to visiting their three children and their growing families.

www.ingramcontent.com/pod-product-compliance
Lightning Source LLC
Chambersburg PA
CBHW031825170526
45157CB00001B/182